Environmental Awareness

Environmental Awareness

Annette Bolder

Oxford Book Company
Jaipur ▪ New Delhi

Published By

Oxford Book Company

Regd. Off. "Bony Residency", Gate No. 2
Opp. Tilak Public School, Vishweswariya Nagar,
Gopalpura Road, Jaipur - 302018, India
Ph: 0141 - 2761280, Telefax: 2761381
email - oxfordbookjpr@gmail.com
website - oxfordbookcompany.com

Delhi Office: 102, 1st Floor, Satyam House,
4327/3, Ansari Road, Darya Ganj,
New Delhi - 110002, India
Ph: 011-45652440

Published in India, 2025

ISBN : 9789380179469

Price : ₹ 2250/-

Typeset by Shivangi Computers, Jaipur

Printed and bound in India by Thomson Press Ltd.,
New Delhi

Preface

Environmental education seeks to make pupils fully aware of the problems connected with the environment and motivates them to tackle those problems with a sense of responsibility and with the technical skills that will enable them to contribute to the solutions for those problems. It is not just a subject of education but an expansion of its whole philosophy recognizing our environment as continuous without shelves and in need of the same care and understanding we give to our personal and social well being. It refers to the awareness of physical environment and its relevance to real life situation, the problems and issues are to be identified, and the imbalances of environment are to be improved in view of sustainable development.

Understanding the very importance of the environmental awareness among students, this study has been taken up. The students are having high environmental awareness. The locality of universities and medium of instruction have no bearing on environmental awareness, but gender of students and management of school have a significant influence on the environmental awareness of students.

This study will be of great use to the teachers and students to take care of the components concerned to the environmental awareness of students.

Annette Bolger

Contents

Chapter 1

Introduction to Geology

Geology is the science of the earth in all its aspects, except those which deal with the relationship of the earth to other planets and to the sun of our solar system. That aspect of the earth properly belongs to the science of Astronomy, for our earth is one of the heavenly bodies with which that science is concerned. It is true that this phase of the subject is often treated by the geologist under the name of *astronomical geology*, but in this book we shall consider it as belonging primarily to the field of astronomy.

At the same time we shall recognize the importance to the student of at least a general understanding of these astronomical relations of the earth, and consider such an understanding a desirable preliminary preparation. The term Geology is derived from the Greek words *ge*, science, or logical discourse. The science of geology was preceded by and in a measure grew out of the subject of geography, which literally is the description of the earth and was, of course, chiefly confined to a description of the earth's surface features, and their significance in terms of human existence.

But since surface features are generally the expression of underlying structure, and of the geological history of the region, it is evident that geography cannot be divorced from geology, and that, to no inconsiderable extent, the student of the geography of a region must take account of its geological structure and history.

SCOPE OF GEOLOGY

In its broadest sense, then, geology is the science of the earth and all that pertains to it. It is the study in — detail of one of the planets in one solar system, by the inhabitants of that planet, who are themselves a part of it. If other planets of our solar system or those of other solar systems are inhabited, the study of those planets by their inhabitants would correspond to our geology — it would be the science of a particular planet.

By us little can be ascertained regarding the structure and

development of other planets, except in an indirect way, — and all investigations along such lines fall properly into the domain of the astronomer, though he concerns himself primarily with interrelations of the heavenly bodies. And thus we may consider that the study of the physical universe, *i.e.*, the *cosmos*, which may be called the science of cosmology, has only the two primary divisions, astronomy and geology. This may be summarized as follows:

- *Astronomy*: The science of all the heavenly bodies,including the earth, their character, distribution, interrelations, movements, etc., and the laws which govern them.
- *Geology*: The science which deals with the material,structure, history, etc., of one of those bodies, i.e., the earth.

Such a view of the science of geology gives it an extremely comprehensive scope, and we must recognize that in ordinary parlance the term is used in a much more restricted sense Nevertheless it is desirable to take this comprehensive view at the outset, and to note the several subdivisions into which such a broad science naturally falls, and with all of which the student of any one division should have at least a general acquaintance. We shall best get the proper view-point by first considering the earth in its entirety.

THE EARTH VIEWED AS A WHOLE

Could we view the earth in its entirety from some extraterrestrial vantage point, we would recognize it as an oblate spheroid, that is, a body approaching that of a sphere, but with its polar diameter flattened and its equatorial belt somewhat inflated. By measurement the polar diameter of the earth is found to be 7899.7 miles, while the equatorial diameter is 7926.5 miles. If the observer in extraterrestrial space is sufficiently removed from the earth's surface, that surface would appear essentially smooth, as does the surface of the moon to our unaided eyes; the irregularities which the dweller on the earth recognizes as mountains and valleys would become of insignificant proportions. Were we to represent the earth by an accurately scaled model 10 feet in polar diameter, the equatorial diameter would exceed the polar by only a trifle over 4/10 of an inch, while the highest mountains on the earth's surface would form elevations on the model less than 3/32 of an inch in height. Thus viewed, the prominences which appear to us formidable are after all of minor significance, and bearing this in mind, we can understand that relatively slight bulgings or crumplings of the earth's surface may produce what to us appear as great elevations.

The wrinklings on the skin of a drying apple constitute far more prominent elevations in proportion to the size of the apple than do the highest mountain chains on the earth's surface, when compared with the size of the earth, while the scratches and notchings formed upon the surface of a glass marble, after a brief play, form vastly greater depressions, in proportion, than do the largest river cañons, such as that of the Colorado, or the deepest ocean depressions.

Thus when the geologist finds evidence that the summit of the high peaks of the Himalaya Mountains or the top of the Apennine chain were once a part of the sea-bottom, he no longer concludes that the sea once covered these mountains, as was formerly supposed, but he infers rightly that these mountain chains were raised by a wrinkling of the earth's crust or by an upward warping which occurred at a time subsequent to that in which this region was beneath the sea.

The evidence from the material of the mountain which leads the geologist to such a conclusion, and the corroborative evidence from the structure of the mountains will also be given. The chief lesson which it is intended to impress upon the student at present is that surface features are relatively unimportant when we consider the earth as a whole, and that elevations and depressions — except those which we recognize as continental masses and oceanic depressions — are of local significance chiefly, and may change from one to the other not once but many times.

The Spheres Open to Direct Study

The observer from his extraterrestrial view-point will, however, note clearly that a large part of the earth's surface — to be precise, about 70.8% of it — is covered with water. This is the sea which is divided into a number of oceans and surrounds all the lands, and, moreover, forms a continuous body, the surface of which comprises approximately 137,070,000 square miles or 361.1 million square kilometers. The land has a surface area of about 59,870,000 square miles or 148.8 million square kilometers, giving a total surface area for the earth of 196,940,000 square miles or 509.9 million square kilometers. Certain portions of the land surfaces are also covered by water bodies, the lakes, which are not in direct connection with the sea. Among these the Great Fresh-water Lakes of North America, and the Caspian, a great salt-water lake, are the most prominent examples.

These are to be classed as a part of the land, as continental water bodies, distinct from the sea. Moreover, the upper layers of the land in nearly all parts are water-bearing, this water filling the empty spaces of the soil, and occupying the pore-spaces within the solid rock. This is the *ground water*, which is tapped by wells, mines, or borings, or issues on the surface in springs which feed the brooks or rivers, or form swamps, ponds, and lakes. It is thus possible to speak of a nearly or quite continuous mantle of water which completely covers some parts of the solid surface of the earth and more or less completely saturates the exposed parts. This mantle of water may be viewed as an aqueous sphere enveloping the rock sphere of the earth and it is spoken of as the hydrosphere. It in turn is enveloped by the sphere of gas and vapour, the well-known *atmosphere.*

These two spheres or shells constitute the outer layers of the earth thus viewed in its entirety, and surround the more rigid mass of the earth, with which we are most familiar. This consists of solid rock, and of unconsolidated soil and other loose material, all of which, except the surface film of organic matter, we may recognize, on examination with a magnifier or by other means, as brokendown rock material in a fine state of division. From observation of soundings, dredgings, etc., and by inference, we know that similar material forms the ocean floor, and thus we recognize that beneath the hydrosphere is a sphere or a shell of rock material.

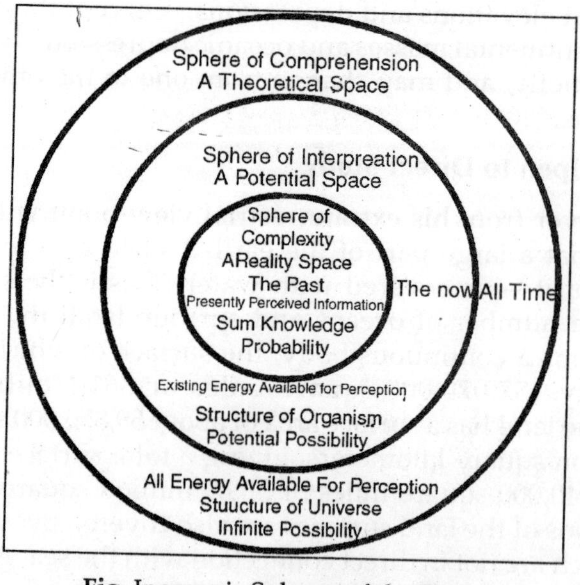

Fig. Inorganic Spheres of the Earth

This is called the rock sphere, or *lithosphere,* and in it are included all of the soil and other unconsolidated rock material of the earth's surface. How thick this shell of rock is, we have no means of knowing from observation. The deepest borings into the earth's crust have penetrated only a little over a mile in depth, while the deeper mines go less than two miles beneath the surface. But from logical deductions of many observed physical facts, we conclude that this shell has a thickness of at least 75 miles, and perhaps much more. It is indeed generally held that the earth is solid rock to the core, but there are those who have held, and some who still hold, that the central portion of the earth consists of fluid or perhaps even of. gaseous matter.

There are thus three inorganic spheres, the *atmosphere,* the hydrosphere and the lithosphere, open to partial observation.

The Inner Spheres

From the observation of many phenomena, geologists and physicists have reasoned that beneath the lithosphere, other spheres characterized by certain peculiarities may be recognized, though their boundaries are variable and probably not very definite. One of these is the sphere or zone in which the temperature of the earth is sufficiently high to permit the fusion of rock if the pressure, which ordinarily keeps it solid by raising the fusing point, were removed or lowered by some structural change in the earth's crust.

There may be regions in which essentially permanent pools of molten rock exist within the crust, forming feeders of volcanoes, but the majority of such feeding areas are more probably temporary. It is convenient to speak of this more or less concentric zone as a distinct sphere and the name *pyrosphere* has been applied to it.

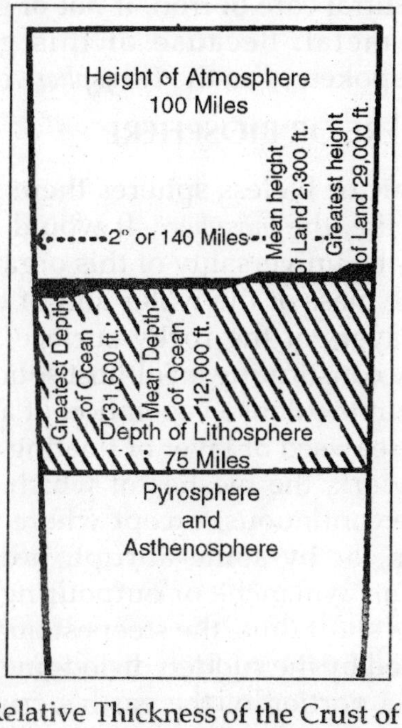

Fig. The Relative Thickness of the Crust of the Earth

Another equally indefinite sphere or zone, variously regarded as lying from 30 to 75 miles beneath the surface (according to our estimate of the thickness of the lithosphere), is one of relative weakness, between a strong external crust and a rigid central mass. Here the rock yields most readily under longenduring strains of limited magnitude, and earth-movements, resulting in the deformation of the rocks, occur. This zone or sphere, which has been called the *tectosphere* or the *asthenosphere*, may in part include the pyrosphere; nevertheless, it is convenient to speak of it as distinct. Finally there remains the central portion of the earth, the *centrosphere*, which is by far the largest part of it, and about which we know nothing from observation and but little from inference. From the known rate of temperature increase in borings and mines, it is inferred that the temperature near the centre of the earth may be between 200,000 and 350,000 degrees Fahrenheit, a temperature so high that were it not for the enormous pressure, all the rocks there would probably be vaporized.

From the fact that the average specific gravity of the rocks which constitute the earth's crust is only about 2.6 while that of the earth as a whole is 5.6, it is further inferred that the material of the centrosphere is very heavy and it has been calculated that if there were a steady increase in the specific gravity of the material from the surface toward the centre, this would, at the latter point, become 11.2, which suggests the possibility of a central core of iron, if not of gold and platinum or some other heavy metal. Because of this greater weight, the centrosphere is also spoken of as the *barysphere* or heavy sphere.

THE ORGANIC SPHERE OR BIOSPHERE

To these inorganic or lifeless spheres there is added the organic sphere, or sphere of life, the *biosphere*. It would surprise the average student could he see the universality of this organic shell of the earth. In and out among the mass of the water and of the air and upon and through the upper layers of the rocky surface, the thread of living matter is woven, now constituting an almost solid mass of tissue, as in the bodies of peat or in the dense vegetation of a forest, or the almost continuous tangle of seaweed or layer of floating organisms in the sea; again forming a network, the meshes of which vary greatly in size, but are on the whole continuous, except where momentarily broken by the hand of man, or by some abrupt, not to say cataclysmic disturbance, such as an avalanche or outpouring of a mass of lava.

But left for even a short time, the steepest quarry wall, or the most precipitous cliff formed by the sudden dislodgment of a mountain side or by the sinking of a portion of the earth's crust, such as may take place during an earthquake-producing disturbance, will be again taken possession of by some form of plant, if not of animal life. Even the stony lava field will be covered in time by a succession of organic forms of increasing complexity. The apparently barren desert, too, has its wide-meshed net of living beings, except perhaps where the sands are constantly in motion; and the presence of these organisms is often shown by the countless tracks and trails which appear upon the surface of the sand on a dewy morning, or the sudden springing up of vegetation from hidden seeds when moisture furnishes the condition for expansive growth. And in the sea, life of some form is never wanting long, — here covering the bottom with a continuous living carpet, there forming an ever changing web of floating animal and plant life, through which the swimming world of animals weaves an intricate pattern with the threads of its never ceasing wanderings.

Thus it is perfectly in accord with the facts, when we speak of a shell or sphere of living matter, the *biosphere,* which surrounds the lithosphere and penetrates its upper layers as well as the hydrosphere and the atmosphere, but is distinct from all. That such a biosphere has existed throughout most of the past ages of the earth's history is clearly shown by the countless remains of the hard parts of animals, such as the shells of mollusks, the bones of vertebrates and the like, which fill many of the rocky layers of the lithosphere, and indeed sometimes actually compose these layers. For example, the white chalk of the English and French coastal regions is literally made up of the shells, and fragments of shells, of organisms (Foraminifera, coccoliths, etc.), and many of our great limestones are consolidated remains of shells of animals which formerly inhabited the seas, or fresh-water lakes. An enlargement of a group of minute needle-like shells, each one of which was built by a separate marine animal and of which a limestone, traceable over wide areas in the state of New York, is almost entirely composed. From careful study, it has been determined that a cubic inch of this rock, when pure, contains about 40,000 individual shells.

Fig. Styliolina Fissurellal

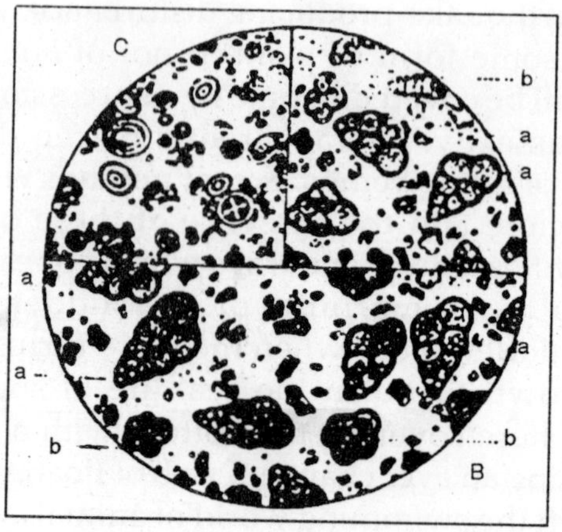

Fig. Thin Sections of Chalk

Whole mountains are sometimes made up of rock, which is largely, if not entirely, the product of accumulation of the hard structures of former organisms. An example of this is found in the famous Dolomites of the Tyrol, the chief rock masses of which are made up of calcareous matter separated from a sea by marine plants, and the many centuries of time during which the present forms of these peaks were carved from this rock have witnessed only a partial obliteration of the organic structures, though no inconsiderable portion of the mass has been removed.

The remains of animals and plants which compose, or are included in, the rocks of the earth's crust, are known to the geologist as *fossils* (from the Latin *fossilis*, something that is dug up) and their study forms an important and integral part of the science of geology. It is indeed the study of the constantly changing elements of the biosphere, of which the existing animals and plants form only the most modern phase, that has made possible the deciphering of the history of the earth.

Beds of coal, too, point to the former extensive accumulation of vegetable material, and the oil of our oil shales and pools owes its origin largely, and perhaps altogether, to the distillation, during many centuries of time, of buried organic matter which constituted a portion of the biosphere of former geological periods.

LITHOSPHERE

We now see that the biosphere contributes no inconsiderable portion to the growing lithosphere, and this is true to a certain extent also of several of the other spheres. For not only was the carbon of ancient plants, which is now locked up in the earth's crust as coal and oil, contributed by the atmosphere of the past, but the water vapour of the air, freezing and descending as snow or hail, or crystallizing directly upon cold surfaces, forms a not unimportant, though in most regions very transitory, addition to the solid rock of the earth. So, too, the conversion of water to ice forms rock at low temperature, for ice is a rock.

More permanent contributions are made by the crystallizing of various salts, chief among them the common salt (sodium chloride) may become buried by other rock material, and be preserved for millions of years as an essential part of the earth's crust. Thus the salt now mined in New York State and in southeastern Michigan was derived from the waters of an ancient sea, the derivation being by a roundabout process.

That this occurred in a period of time many millions of years ago is amply shown by the ascertainable facts. So, too, the great salt deposits of North Germany, which have furnished the world in the past with most of its potash, are the product of the evaporation of cut-off portions of an ancient sea which covered much of Europe, though the period of its production is not quite so remote as that of the best-known American salts.

That the pyrosphere also contributes its quota to the rock mass of the earth is abundantly shown by the vast extent of ancient lavas and other igneous masses which now form a large proportion of the solid crust of the earth and are visible to us as the result of surface cooling in comparatively modern times, as in the case of the great Columbia lava fields of Northwestern America and the preserved fragments of similar flows on the British coast, or which were uncovered by the removal, during long geological epochs, of the overlying portions of the earth's crust beneath which the ancient igneous masses assumed their solid form.

Throughout the entire history of the earth, the lithosphere has received additions by the cooling of molten rock material injected into it or poured out on its surface, and this addition is in progress even to-day.

SCIENCE OF GEOLOGY

IT is but natural that the mind of man, always eager to enter into details, should have developed distinct branches of inquiry into the character and history of the several parts of the earth which is his home. In the beginning of such inquiry, a single master mind may have encompassed the entire field; but with the increasing mass of detail, the individual range became limited to special portions of the field, and in the course of time the several lines of inquiry into the nature and history of our earth developed into distinct sciences, each with its host of devoted searchers after truth.

We may, then, at this point of our study, endeavor to ascertain the natural subdivisions of the Science of the Earth, or Geology in its broadest sense, and after that confine ourselves to those branches which by common consent are reserved as the special field of the geologist, as it is understood at the present time.

Atmology or Meteorology

From an inspection of the table of spheres into which the earth may be divided, we form the natural conclusion that the first line of cleavage would be in accord with these natural subdivisions. The study of the atmosphere has now developed into a separate science, which is familiarly known as *Meteorology* because the meteors, or extraterrestrial bodies, the familiar "shooting stars," which on entering our atmosphere become luminous through friction and frequently reach the earth as meteorites, have from the remotest days formed one of the chief attractions for those whose gaze was turned away from the solid earth. Thus atmospheric phenomena came to be designated as meteoric phenomena and the science of these phenomena became meteorology. But since the sphere with which this science is concerned is the *atmosphere,* a more appropriate name for this science itself is *Atmology.*

Hydrology

That the hydrosphere, that is, the oceans and other water bodies, early attracted the serious-minded student was but natural, since these water bodies form the natural highways of commerce, and since they furnish so large a part of the earth's population with the means of a livelihood. Thus the science of the water, or *Hydrology,* was developed, which is variously subdivided again into the sciences of the oceans, or *Oceanography,* and into the sciences of the lakes, the ponds, and the rivers.

That some of these, for example oceanography, have been developed to a remarkable extent, is shown by the great oceanographic institutes of the world, of which that of Berlin under Director Penck, and that at Paris, under the direction of the Prince of Monaco, are the best known. In America, oceanographic studies have chiefly been carried on by the federal government and by private individuals, among these the late Alexander Agassiz; and there is a hydrographic department of our government, the chief business of which is the charting of our sea-coasts, lakes and rivers.

In England, too, oceanographic studies have been largely fostered by the government, under whose auspices the famous Challenger Expedition, for the exploration of the oceans, under the leadership of Sir John Murray, was carried to successful completion. The governments of other countries, too, have fostered exploring expeditions for oceanographic research, these being generally known by the name of the vessel employed in the expedition.

The student of geology, even in its narrower sense, must keep abreast of these inquiries, at least to a moderate extent, for without them many of his discoveries lack the clarifying light by the aid of which he must interpret them; and in proportion as the researches of the oceanographers and the other hydrologists become available, the analysis of the structure and composition of the earth's crust, the special field of the geologists in the modern sense, and the interpretation of these facts in terms of earth history, become more precise.

Lithology or Geology Proper

As has just been said, the special field of inquiry of the modern geologist is the crust of the earth, or the *lithosphere*. He is thus primarily a student of *Lithology* -though that term has also been used in the past in a narrower sense, the study of rocks *per se*.

It will be well, however, for the sake of unity of terminology — a great desideratum in every science — to adopt this term in its wider sense, namely, that which encompasses all the fields of inquiry which are concerned with the composition, structure, and history of the earth's rocky shell, the *lithosphere*, generally assigned as the only legitimate field of the modern geologist.

Still, to-day more than ever, the geologist cannot limit his inquiries to the facts disclosed by the rocks alone, but must draw his conclusions with the aid of a wide knowledge of the researches of the atmologist, the hydrologist, and even the biologist.

Biology; Palæontology

Biology is the science of living things. It is the study of the *biosphere*, and has perhaps been more sedulously developed, and has a larger body of votaries, than any other branch of the earth science. Biology, as currently understood, limits itself to the study of modern living beings, the plants and animals of to-day, and naturally falls into the two divisions, or separate sciences, of *Botany* (Phytology), the study of plant life, and *Zoology*, the study of animal life.

Biologists are, however, aware of the fact that the modern world of plants and animals is only a fragment of the great host of living beings which has inhabited the earth from the remotest ages. Throughout all the past ages, living forms existed upon the earth, and their remains are embedded in the rocks of the earth's crust as *fossils*. The study of these was at first left to the geologist, but has now developed into a separate science, that of *Palæontology*.

Other Divisions

When we come to the consideration of the inner spheres of the earth, we must admit that, owing to our inability to make direct observations, no separate sciences worthy of that designation have yet been developed. It is true that a science of earth disturbances or *Seismology* exists, and that this properly belongs to the study of the asthenosphere or tectosphere.

But the observations on which this science is based are chiefly possible upon surface manifestations and the effects which these produce, though inferences can also be drawn from the study of disturbances which have occurred in the past, the effects of which are recorded in the rocks. The same is true of the study of the pyrosphere — which can only be approached through observations of modern volcanoes (*Vulcanology*) and of the temperatures in deep borings and mines. Of the interior of the earth we may perhaps never know much through observation, and so deductions from physical phenomena alone must guide us. No separate science of the centro- or barysphere has thus been developed.

From the foregoing, the student is led to the realization of the truth that all true sciences are based on two great fundamentals, *observation* and *deduction*. To observe and record facts alone does not constitute the whole of the work of the man of science. Such used to be the sole aim of the older naturalists, so called, who dissented from the attitude of the ancient philosophers, because these based their speculations on mental processes rather than observation of facts. To-day, however, we know that the interpretation of facts, in the light of our growing knowledge of causes, is as legitimate, indeed as important, a function of the student of the earth as the observation of facts, and moreover a certain amount of speculation — always held in check by the appeal to facts — is not only desirable but necessary. The student of nature must be a philosopher in the true sense of the word, and so long as he does not lose sight of his fundamental base — the observation of facts -his work will gain in value by allowing his reasoning faculties their fullest play.

LITHOSPHERE

Restricting our attention now to the more limited field to which the modern geologist applies himself, namely the study of the material, structure, and history of the earth's crust, or the lithosphere, we may note at the outset that there are two phases or aspects of this field, one or the other of which is cultivated more sedulously by the geologist according to his predilections or circumstances, but neither of which can be wholly neglected by the worker in either field.

These two aspects are the purely scientific, and the practical or applied. The geologist who works in the pure science field of his domain does so primarily for the intellectual satisfaction derived from the discovery of facts and principles. His aim is chiefly to search for nature's truths irrespective of their bearing on human welfare, and his principal endeavor is directed toward widening the boundaries of human knowledge and pushing forward the frontiers of discovery.

The chief aim of the practical geologist, on the other hand, is directed toward making the forces and materials of the world available to man, to augment the welfare of the human race, and to push forward the boundaries of civilization.

But this work, important and noble as it is, can only be carried on successfully in proportion as the facts and laws of the science are discovered and demonstrated, and the practical geologist must forever depend on the worker in pure science, whose reward too often is little more than the satisfaction which is to be derived from a devotion to the search for truth. We shall for the sake of convenience speak of the pure-science phase of our subject as *the scientific aspect of geology*, and of the applied phase of geology as *geology in its relation to man*. Each phase has its special subdivisions, to which attention may now be drawn.

THE SCIENTIFIC ASPECT OF GEOLOGY

Structural Geology

A logical analysis of this field of investigation reveals a threefold aspect and three corresponding methods of approach in study. In the first place the material of the earth's crust, and the composition and the structure of this material, invite attention.

This portion of the subject is generally treated under the caption *Structural Geology*. It takes account of the composition, form, and architecture of the earth's crust, and is primarily an analytic branch of the science. From this point of view the earth's crust may be considered under the following subdivisions, given in the order of their magnitude.

Subdivisions of Structural Geology

These include the following fields:

- Chemical elements and ions.
- The combination of these elements and ions into salts and other compounds which either take on crystalline form or remain in an uncrystalline (amorphous) condition. These compounds as they occur in nature are designated minerals.
- The combination of crystals or fragments of the same or different minerals into large masses, either bound together into a more or less solid mass, or remaining in an unconsolidated condition. These are the *rocks* of the earth's crust, and for their study a recognition of at least the principal component minerals is essential.
- The association of rock masses into original or primary *structural units,* which constitutes the primary *architecture* of the earth's crust; or the impression upon it of *secondary structures,* through deformation or other influences of an outside nature. This is structural geology in the narrower sense — and has also been called *Geotectology*. The original structure corresponds to the initial architecture of a building, the stones or bricks of which correspond to the rock masses of the earth's crust. The secondary structures correspond to the changes subsequently made by additions, removal or modification resulting from sagging with age, etc.
- The surface forms resulting through the activities of destructive as well as reconstructive forces, and recognizable in the mountains and valleys, the plains and plateaus, and other physical features, generally considered under the subject of *Physical Geography*.

The term *Geomorphology* or the study of the details of the earth's surface forms has been commonly applied to this branch of inquiry, when it is not merely descriptive or analytic, but takes account of the history or genesis of the form as such. It corresponds to the study of the form of a building as a whole, either in its completeness or when changed to a ruin by destructive influences.

The study of most of these divisions of structural geology has been carried to such detail that separate sciences have been developed. Thus the study of the natural substances, or minerals, has become the science of *Mineralogy,* and that of the rocks the science of *Petrology* (also sometimes designated as lithology in its narrower sense), while the science of rock structures is *Structural Geology* in the narrower sense, and that of the surface forms is *Physiography,* and their relation to man is *Geography*.

In all such studies, while emphasis is laid on the analytical and descriptive sides, the causal or dynamic side and the historical or developmental side are given their due attention. This is expressed by the preference of terms ending in *ology* (mineralogy, petrology, geotectology, geomorphology) over those ending in *graphy* (petrography, physical geography, etc.), which emphasize mainly the descriptive and analytical side.

Dynamical Geology

A second mode of approach is that which lays the emphasis upon the forces that work upon and within the earth to produce results, which, in this view, take a secondary rank. This is *Dynamical Geology* and it naturally falls into the two great divisions, the *physical* and *chemical.*

The sciences of physics and chemistry, in so far as they lay the stress upon the forces at work, are the specialized development of these aspects. That chemistry, in its analytical as well as its dynamic aspects, is of fundamental importance to the geologist has generally been recognized, but the importance of physics to the student of earth science has only recently received proper recognition by the establishment of geophysical laboratories.

Both *chemical* and *physical* forces may be viewed as *constructional,* or those building up rock masses and rock structures, and as *destructional,* or those destroying them. A third view of these forces is that which deals with their effects in modifying or deforming the materials and structures, and this may be termed the *reconstructional* or deformational aspect.

Geology

The third method of approach is the historical or evolutionary method, in which emphasis is laid upon development and the causes which underlie this development. It is apparent that this phase of geology is the latest and most specialized aspect, and that for its proper prosecution a thorough preparation in the other two phases is needed. Moreover, since the history of life upon the earth is intimately bound up with the history of the lithosphere, a knowledge of biology, and especially of palæontology, is indispensable for the prosecution of any but the most general studies in earth history.

Several special branches have been developed within this field. One of these deals with the origin or genesis of the rocks and their structures, or the origin of the lithosphere. To this branch the name *Lithogenesis* is commonly applied. Another branch considers the character, arrangement, succession, order of formation, and age relations of the stratified rocks. This is the science of *Stratigraphy*, primarily a descriptive one.

A third branch deals with the succession and distribution of the organic remains, the petrifactions or fossils in the rocks in so far as they have a bearing on the geological history of the earth, *i.e.*, the *index fossils*. This is the special geological phase of *Palæontology*, which has also been designated by the name *Petrifactology*. Again, there is the science which deals with the development or genesis of the surface forms of the earth not merely in the descriptive manner, but from the view-point of origin.

This, as already noted, is *Geomorphology* in its proper sense, though it is more commonly known by the name of *Physiography*, which formerly had a very different meaning. Finally there is the study of the changes in the earth's surface, or its geography through the successive ages of the earth's history, — together with the changes in climate and the dispersions and migrations of the organisms and the causes which effected these. This is the latest of the several aspects of Historical Geology and is now termed *Palæogeography* or the geography of the past.

In this field the palæogeography of the Pleistocene period has been most extensively studied, and a separate branch, that of *glacial geology*, has been developed. *Geography*, in the usual sense of the term, is the geography of the modern or Holocene period of the earth's history.

GEOLOGY IN ITS RELATION TO THE WELFARE OF MAN

So far we have been considering geology in its pure science aspect, that which appeals to the inquiring mind of man in search after truth and knowledge, without ulterior motives of usefulness. There is, however, another aspect of our science, and one which in recent times has come strongly to the front. This is *applied geology*, in which geological facts and forces are viewed in their relation to the needs and requirements of man.

As has already been intimated, the application of any science for any purpose whatsoever is successful in direct proportion to the profundity of knowledge possessed by the applier. No successful exploration of geological products or application of forces is possible without a thorough understanding of the facts and principles of the science, and the student who wishes to follow the applied side of his science should not fail to make his preparation in the pure science side as broad and as profound as circumstances will permit.

Among the earliest problems, to the solution of which geological knowledge has been applied, are those of mining. Indeed, the science of geology, in a measure, developed in response to the needs of the miner for accurate knowledge of the conditions of occurrence, distribution, and mode of origin of the valuable mineral deposits. To such an extent has this been carried, that a separate branch of *mining geology* has come into existence.

Moreover, as our knowledge increased and the possibility of more detailed application of our science became apparent, special subdivisions of mining geology have been developed, and it is found that individuals can profitably devote themselves to the cultivation of a narrow field, to the practical exclusion of the others. Thus there have been developed the branches of *coal geology*, of *petroleum* and *gas geology*, and of *salt geology*, including the geology of potash, phosphates, nitrates, borax, and other salts, and of bauxites and other aluminum ores.

In these branches the investigator confines himself to the problems involved in the occurrence of these substances, which are chiefly restricted to the stratified rocks. It is now well recognized that successful search for such deposits can only be undertaken by one well versed in the science of stratigraphy (including index fossils) and structural geology, while a thorough understanding of the principles of physiography and palæogeography is almost indispensable.

The mining geologist who devotes himself primarily to the problems of the metallic deposits must have not only a thorough knowledge of mineralogy, petrology, and structural geology, but of dynamic geology as well, and especially of the chemical and physical principles involved in ore deposition. As many ores are also found in stratified deposits, a knowledge of stratigraphy and of index fossils is necessary, while an understanding of the principles of physiography and palæogeography will also be found of value in many cases.

Geology, too, plays an indispensable rôle in the solution of many important engineering problems. In the construction of the Catskill aqueduct for New York City, a force of competent geologists was constantly employed, and specialists were frequently called upon for consultation. This same need was felt after the construction of the Panama Canal had been undertaken, and a resident geologist was appointed to supervise the later phases of construction. That many difficulties might have been avoided had such supervision existed from the outset of the undertaking, is now generally conceded.

Geological advice has always been employed in the construction of great tunnels such as those piercing mountains or passing under rivers and other water bodies. In many cases, too, the selection of sites for bridges, dams, and other great engineering works has been based on geological advice, while in other cases, where such advice has not been sought or has been disregarded, disastrous consequences have resulted. In consequence of the growing recognition of the need of geological study in the undertaking of engineering problems, the special branch of *engineering geology* has been developed.

The geological engineer must be primarily a structural geologist and one who has a thorough grasp of the principles of dynamic geology, including hydrology as well, while physiography too is of great importance to him. To a lesser degree a knowledge of rock types, of stratigraphic principles, and of index fossils will be needed by him, and not infrequently a knowledge of palæogeography, especially that phase which deals with the Pleistocene, or the problems of the glacial period, will be found of the greatest value.

Finally there has been developed in recent years the special branch of geology which deals with the problems involved in military campaigns, and to this the name *war* or *military geology* has been applied. Some of these problems are concerned with the proper location of sites for camps, and trenches, and with water supply and sanitation, and for these a knowledge of structural geology, of rock types, of stratigraphy, and of glacial geology has been found necessary.

Other problems are of an engineering type and require the preparation of the geological engineer. Again, the problems involved in military operations need for their solution a well-trained physiographer and a competent meteorologist as well. Problems concerned with naval warfare require the attention of one well versed in hydrology, especially that phase of it which deals with the oceans, or oceanography.

There are other ways in which a knowledge of geology has become useful to man, and as the science itself is developed new channels of application into which it may be directed will no doubt be discovered.

THE HISTORY OF GEOLOGY

The student should further realise that the development of his science, the history of geologic thought, cannot be neglected by him. We profit by the mistakes of our predecessors as much as we do by their achievements, and the history of the discovery of facts and of the development of geological opinion since the days of the Greek philosophers is fraught with lessons equal in import to those gained from the pursuit of the history of any other department of human thought and endeavor.

At this point it will be desirable for the student to read the masterly sketch of this history from the pen of Sir Archibald Geikie, the book entitled *Founders of Geology*, and if possible follow this by a perusal of the older historical sketch by Sir Charles Lyell in volume I of his *Principles of Geology*. For greater details the student is finally referred to the *History of Geology and Palæontology* by Carl von Zittel, translated into English by Maria Ogilvie Gordon. The history of geology in America is adequately and fully treated by Dr. George P. Merrill in his book, *Contributions to the History of American Geology*.

STUDY OF THE EARTH

The geologist is, above all things, an observer in the great outofdoor world. The man whose horizon is bounded by the walls of a city can never be a geologist, though he may gain much scientific knowledge from books and from an inspection of collections in museums and laboratories.

The true geologist, however, goes directly to the earth and there begins his inquiries. Not until observations of natural facts and phenomena were made in extenso was the inquiry of the philosophers regarding the earth and its history placed on a scientific basis. Scattered observations and more or less accurate deductions were made even in antiquity.

Thus Aristotle, in the third century B.C., had a very considerable understanding of the work of rivers and reasoned correctly regarding the changes in the land and sea at former times. The painter Leonardo da Vinci correctly reasoned, from an observation of the fossils found in the foothills of the Apennine Mountains, that they were the shells of once living animals, though they were generally regarded either as freaks of nature or as modern shells dropped by the pilgrims in their voyages across these mountains.

It is true that the significance of fossil sea shells was recognized by the Greek philosophers but their explanations were generally ridiculed during the Middle Ages. It was, however, not until the latter part of the eighteenth and the early part of the nineteenth century that systematic investigations of the rocks of the earth and their contained fossils began, and from this period we date the birth of geology as a science. Scientific geology arose more or less simultaneously in the different countries of Europe. In France Etienne Guettard and Nicholas Desmarest were among the first to bring observation of facts to the fore, while Buffon indulged in brilliant speculations on the origin of the earth.

Later Alexander Brongniart investigated the rocks around Paris, and Georges Cuvier described their fossils. In Germany Abraham Gottlob Werner, who is often called the founder of German geology and who was Proessor at the Mining School at Freiberg i. S., exerted a profound influence on geology especially by his teachings, to which men flocked from all countries.

Being mostly an observer of the details of specimens and rarely venturing beyond his immediate surroundings for field observations, many of his geological deductions have proved erroneous — though his pupils and followers, notably Leopold von Buch, extended their observations over wide areas and added much to the store of geological facts as well as to its philosophy. Switzerland had its enthusiasticstudent of the structure and history of the Alps in the person of H. B. de Saussure, and Russia had in Pierre Simon Pallas its careful student of the Ural Mountains, and the rocks outcropping there and elsewhere in the empire.

In Great Britain many men contributed to the discovery of facts and the interpretation of these. Among them the first rank is given to James Hutton, whose work marks a turning point in the history of geology, for he insisted that "the past history of the globe must be explained by what can be seen happening now, or to have happened only recently," a dictum which has since become the very cornerstone of geology.

Hutton great work, *Theory of the Earth with Proofs and Illustrations,* is better known through the classic volume, *Illustrations of the Huttonian Theory,* by his friend John Playfair, which no student of geology should neglect to read. In this work are contained many of the fundamental principles with which geologists are concerned to-day, and they are illustrated by a wealth of facts gleaned by Hutton from his rambles through Scotland and other countries.

Another of the Scottish founders of geology was Sir James Hall, to whom we owe the origination of experimental geology. The best known, however, among the early English geologists was William Smith, who is generally called the "Father of English Geology." He determined not only the correct successions of the English rock formations, and made the first geological map of England, but gave to many of the formations the names which they bear to-day. Finally, the student should remember among English geologists the name of Sir Charles Lyell, as that of a man who has had the most profound influence on geological thought. His great work, the *Principles of Geology,* has become a classic of geological literature.

Among the men who exerted a profound influence on American geology in the early days of its development, the names of William McClure and Amos Eaton stand out prominently. McClure, born in Scotland in 1763, became an American citizen near the close of the century. In 1809 he published the first important work on American

Geology, in which appeared the first geological map of the Eastern United States, and one of the first geological maps of the country. Amos Eaton, born in New York state, is known especially for his *Index to the Geology of the Northern States,* which was the first geological textbook published in America, and in this and subsequent works he laid the foundation for the New York geological system.

Many of the names of American formations still current were first applied by him. He also made the first geological map of New York. The important influence which McClure and Eaton had on the development of American geology has been recognized by the designation of the first two eras in the history of this science in America as the Maclurean and the Eatonian.

THE FIELD OF GEOLOGICAL OBSERVATIONS

"Where, then," the student will ask, "can the facts of geology be observed, and how have they become available?" To get at the facts of structural geology the student must go to the rocks. True, the rocks and the minerals and the fossils are brought to him in museums and laboratories, and he will do well to begin his studies of selected examples thus brought together and capable of being examined under the most favorable conditions.

But the knowledge thus gained must be amplified and correlated by repeated visits to the home of the rocks, where alone their larger relations and their true significance in the history of the earth can be ascertained.

THE FIELD FOR THE STUDY OF ROCKS AND ROCK STRUCTURES

Rock Exposures in Flat Countries

The dwellers in the interior of our country, or the traveller on the broad plains of northern Germany, of Russia, of Hungary, or of China, will find little opportunity to get a view of the rocks which underlie these regions, for an almost continuous mantle of soil and drift covers the solid rock. Only where rivers have cut channels through the surface layer of loose material, the *mantlerock,* or where quarries have been opened, or construction operations have necessitated excavation down to and into the solid bed rock, is there an opportunity for observation of the rocks beneath the mantle-rock.

River valleys and gorges are therefore the favourite resorts of the geologist of the plains, while quarries, railroad cuts, and other excavations which expose the rock, likewise receive his attention. Borings and drillings for oil, gas, or water often prove of use to him, though generally, except where the diamond drill is used, and the core preserved, the record of such borings is only of doubtful and minor value.

Salt shafts, such as those in central New York and that near Detroit, Mich., furnish excellent sections at the time of excavation. Fortunately for the geologist of the plains, however, few regions of any great extent are wholly covered by mantle-rock; more commonly lowlying ridges of rock, formed by resistant layers, are exposed above the general surface of the soil and drift mantle. Such "ledges" generally mark the edges of beds of hard rock which have a gentle slope away from the edge of the ledge. When the ledge-forming layer is very thick, a cliff of some height may result, and this generally furnishes abundant opportunity for observation and deduction.

All natural exposures of the rock are called *outcrops*, and the outcropping ledges together with the exposures in the stream channels, especially those which cut across the cliffs, furnish to the geologist of the flat countries his most satisfactory data. We may note a few examples.

Illustration from New York State

The state of New York furnishes an illustration of the types of rock outcrop in ledges referred to in the preceding paragraphs. Over the greater part of the state the rocks are so gently inclined that they appear horizontal to the eye, and it is only when they are seen in the cliffs of Niagara gorge and along Lake Erie that their gentle southward descent becomes noticeable.

Such a section, considerably generalized. Where the beds end in the air upon the north, the harder ones, such as limestones and sandstones, form a series of low cliffs, while the ends of the softer shale beds are usually marked by broad flat-bottomed valleys. The largest and deepest of these valleys is occupied by Lake Ontario, while others are filled by soil which conceals the rock.

These cliffs or escarpments generally have an abrupt northern face across the edge of the hard rock, and in these faces quarries are generally opened. These cliffs can be traced with more or less interruption across New York state from the Niagara River and Lake Erie to the Hudson, although they become modified because some of the beds seen in the western part of the state die out, or change in character and new ones appear.

This is shown on the geological map of the state of New York, where a series of broad colour bands extends east and west across the state. Each colour band in general represents one rock layer or group of layers, and the width of the colour band indicates the amount of exposure of each which would appear were all the covering soil and vegetation removed; or in other words, the amount by which each lower bed projects beyond the next higher one which covers it.

Wherever streams have cut across this series of beds, gorges are formed, in the walls of which the cut edges of the rocks, the soft layers as well as the hard ones, are shown. The most striking examples of such gorges are shown along the Genesee River, which crosses the state from south to north.

In the section from Rochester northward it cuts the lower beds, the harder strata producing waterfalls. In the section between Portage and Mt. Morris, it cuts the higher strata, and here too several waterfalls are formed by hard layers. Between these two points many smaller tributary streams have cut into the sides of the valley and exposed the rocks. Here, too, are situated several deep shafts which go down vertically to the Salina salt beds and during the cutting of which the succession of the rocky beds was ascertained.

A good understanding of the succession of this series of rocky formations is obtained by the traveller who passes from the Adirondack Mountains southwestward across the state to Elmira, especially if he take advantage of the various sections exposed in the gorges of the streams and the banks of the Finger Lakes.

It was by the study of these natural outcrops of the state, supplemented by those made on the sections exposed during the cutting of the Erie Canal in 1817-1825, that the foundation of American geology was laid by such men as Amos Eaton, and by James Hall and others associated with him on the geological survey of New York state.

Other Exposure of this Type

The type of outcrop just described is found in most regions of flat-lying rocks in our own country as well as abroad. Thus the geologist who starts from Bamboo, Wisconsin, and proceeds southeastward to Lake Michigan at Milwaukee, will cross such a series of rocky ridges separated by soilfilled valleys, and a similar experience awaits the traveller across Kansas from southeast to northwest, or the one who proceeds southward across Oklahoma, or journeys from central Texas either northwestward or southeastward.

The traveller across England from Liverpool to London also crosses such a series of rocky ridges, which in general extend in a northeast-southwest direction, and are formed by a succession of nearly horizontal rock-layers with westward facing cliffs, though these are not generally visible from the train. Similar outcrops of nearly horizontal layers of rock surround Paris in constantly widening circles on three sides. The edges of some of these rocks form cliffs or escarpments facing outward.

If one were to represent the rock formations which underlie Paris, and which have a shallow basin-like structure, by a nest of plates, the smallest at the top, the successive rims of these plates would represent the encircling cliffs, while the location of Paris would be in the centre of the uppermost and smallest plate. Many rivers have cut channels through the edges of these rock plates, while others flow in the depressions between successive rims.

Thus numerous and excellent exposures of the rocks are found, even though the marginal portions between the edges of successive rock layers are frequently covered by soil and vegetation. These numerous rock exposures made possible the observations which gave the French founders of our science the data on which to build their deductions, and the cliffs which they form around Paris were of the utmost importance in the conduct of the Great War so recently closed.

Outcrops in Mountainous Countries

It is, however, in the elevated portions of the earth — the mountains and chains of uplands — that rock outcrops are most frequent on the surface of the land. Here the soil and rock débris lodges only in the depressions, while between these the ledges protrude and give opportunities for observation. Here, too, deep cañons are cut by the rivers and glaciers and thus additional rock-walls are opened for observation.

German, Austrian, and Swiss geologists have for the most part been limited to such regions for their observations, and the wonderful rock exposures of the Alps and other mountains have made it possible for them to carry their studies along certain lines to great lengths. Werner, the father of German geology, made most of his observations in the subdued mountain district of Saxony, especially the Erzgebirge, which has long been famous for its old and extensive mining operations.

Since France and Italy also border on the Alps, and have mountain ranges of their own, the geologists of these countries were enabled to avail themselves of the rocks and rock structures thus revealed. British geologists, too, have been able to some extent to resort to this type of exposure, though in these moisture-enveloped islands, as over parts of Scandinavia as well, the dense though low cover of vegetation and the peat accumulations obstruct much of the underlying rock, as all students of Irish geology know only too well.

The Highlands of Scotland, however, furnish many good opportunities for observation, as do also many of the higher English districts and especially the mountain region of Wales. Swedish geologists have frequently had to resort to digging through the surface layers to get at the underlying rock, and it is not an uncommon sight to see the Swedish geologist in the southern interior accompanied by a factotum, whose duty it is to wield pick and shovel.

In European Russia and adjoining districts there are vast areas of flat-lying rocks covered by soil and drift, so that, except along the coast and in the river channels, outcrops of the bedrock are difficult to find. But where these rocks are uplifted — in the Ural Mountains on the east, the Caucasus on the south, and the Carpathians on the west — outcrops abound, and here the true relationships of the rock formations may be ascertained.

In America, the New England uplands, the White, Green, and Adirondack Mountains, and the Appalachian Chain furnish an abundant series of rock outcrops for the eastern geologist.

The old and generally much-subdued mountain system, the rocks of which may be traced by frequent outcrops from New York City to the Highlands of the Hudson, and which can be followed southwestward through New Jersey, Pennsylvania, and Maryland, and indeed all the way down to the Carolinas, where it constitutes the older Appalachian Chain, is a typical example of an elevated, though for the most part not very mountainous, country, and here outcrops abound.

Many of our western mountains are especially well adapted for geological observation, for here the aridity of the climate prevents the growth of much vegetation, and the rock structures, frequently developed on a gigantic scale, are visible for many miles. This makes the Rocky Mountains a veritable paradise for the American geologist.

OUTCROPS UPON THE COAST

Of all the natural rock exposures, however, those of the coast-line are the most attractive, and in many respects the most satisfying. Wherever the sea-coast or the shore of a large lake is formed by rocks which rise above the surface of the water, the cutting work of the waves keeps the exposure fresh. A tramp along a rocky sea-coast is replete with interest to the geologist, and many of the choicest bits of geological observation have been made on such sea-cliffs.

Great Britain, with its wonderful rocky sea-coast, probably leads the world in the variety and significance of rocks and rock structures there exposed. No student of geology can afford to neglect the wonderful English and Scottish coast, which has furnished the British geologists so many opportunities for the observation of facts that there, more than elsewhere, geological science has advanced, since the days of William Smith, with phenomenal strides.

This is probably the reason why English geology quickly became the standard of comparison for other nations, in whose home countries observation was a more arduous task, because they did not include such marvelous coast exposures. Northern France, too, has a coast-line of great interest to the geologist, and so has Norway. The coast-line of Germany, on the other hand, is mostly sandy, and there is little diversity in the types of the facts which it discloses. The Atlantic coast-line of North America is for the most part a sandy one. Only in New England, in the Canadian coastal prov inces, and in Newfoundland can be seen coastal sections comparable to some extent to those of Great Britain. The northern regions, however, are accessible with difficulty, and have only recently been investigated. But the New England coast, and especially that of Massachusetts, is a Mecca for American geologists, and many of the workers in American geology have had their preliminary training through a study of that interesting region.

STUDY OF DYNAMIC GEOLOGY

Although the principles of dynamic geology, the workings of the chemical and physical forces, may be studied to much advantage in the laboratory — such study, too, is incomplete without recourse to the outdoor field. It is upon the sea-shore that some of the profoundest lessons of erosion by waves, of transportation by currents, and of deposition in quieter waters can be learned.

Here, too, the method of entombment of fossils and the formation of many original structures, such as ripple-marks and the like, can be observed. Rocky as well as sandy and muddy shores should be visited. Shores of large lakes may serve as a substitute in inland regions, but lakes have in addition many characters of their own. Ponds and temporary pools also teach their lessons.

River valleys and gorges, rapids and waterfalls, brooks, and even the roadside gutter, furnish lessons in dynamic geology, as do also the hillside, the mountain slopes, and the elevated peaks, where rocks are shattered by frost, and decay under atmospheric influence. Glaciers present many illustrations of dynamic geology, while caverns and underground channels have special lessons to teach.

The deserts and all regions where wind is at work furnish illustrations of the mechanical activities of the wind, while pools and salt pans in and regions furnish illustrations of chemical activities and of precipitation of salts through condensation of the water under evaporation. Springs, too, illustrate dynamical activities, both physical and chemical, and artesian wells, oil wells, geysers, and similar phenomena are replete with them. Finally, volcanoes and other such phenomena furnish the means for the study of igneous activities.

The great English geologist, Sir Charles Lyell, whom we sometimes call the " Father of Modern Geology," has said that the geologist must be primarily a traveller — he must go to other lands than his own and so widen the scope of his experience. Werner, the founder of German geology, confined his observations mainly to his limited Saxon district, and attempted to formulate from these observations laws which should govern the rest of the world. Naturally he fell into many and profound errors, so that to-day scarcely one of his theories is held.

Since his day German geologists have, however, become great travelers, not only in their own but in most other lands. As a result, their observations have become of wide scope, and they have added much to geological knowledge. British and American geologists have only recently begun to follow the advice of Lyell, but already their efforts have been crowned with considerable success.

Let the student of geology, then, come to realise that the value of his deductions increases in proportion to the range of his observations, and that no single country or region of the world will give him all he needs. The American student, owing to the wide extent and diversity of his country, is perhaps more favored in this respect than is the geologist of any other nationality, but at present only a limited portion of our country has become sufficiently accessible to make extended observations possible.

THE IMPORTANCE OF GEOLOGICAL LITERATURE

Finally, it must not be overlooked that the observations of our predecessors are recorded in the literature of the science, and that here we find a mine of information, the value of which cannot be overestimated. No one can repeat all of the observations which have been made in the past, even were such repetition desirable. In addition to the laboratory and field, then, the student of geology must go to the library, and a thorough understanding of the literature on his special field is of fundamental importance to the worker.

Besides special books on different aspects of the science, the student should gain familiarity in the use of the official publications issued by the governments of the various countries, the proceedings of scientific societies, and the special journals devoted to geology and kindred sciences.

Chapter 2

Movements in the Atmosphere

The Atmosphere

The atmosphere is an envelope of gas around the earth. At the earth's surface the air contains about 79 per cent nitrogen, 21 per cent oxygen, 0.03 per cent carbon dioxide, small but variable amounts of water vapour, and small amounts of chemically inert gases.

CLIMATES

Word "climate," from the Greek for "incline," preserves one of the oldest scientific discoveries, that year in and year out weather is largely controlled by the tilting of the earth's axis of rotation to the plane of its path around the sun. The Greeks noticed that, in general, the climates are colder at higher latitudes; they correlated this with the obvious fact that the air is generally warmer when the sun is high above the horizon.

As the sun rises higher in summer than in winter (a fact that indicates tilting of the earth's axis), it was natural to attribute seasonal weather differences to this tilting. Modern studies of *weather* (the day-to-day variation) and *climate* (the long-period weather, including seasonal changes) show that many other factors besides tilting of the earth's axis are involved. Weather and climate involve not only heat and cold, but also such features as wind and calm, the direction and intensity of the winds, cloudiness, and precipitation either as rain or snow.

All of these aspects are closely interrelated, and depend on such factors as heat supplied to the atmosphere by the sun's radiation, the movements of the atmosphere, and the distribution of land, sea, and mountains. A full treatment of these interrelations is the subject of a separate earth science-meteorology, the study of the atmosphere—but any geologic study requires some acquaintance with them.

Not only do geologic processes differ greatly in different climates, but the clear evidences of drastic climatic changes in the geologic past cannot be appreciated without some understanding of the factors involved. We begin our study with an analysis of the nature and movements of the atmosphere.

Air Pressure

As in all gases, every portion of the atmosphere exerts pressure against every adjacent portion. Such pressure is one of the evidences that gases are made up of particles in constant motion, colliding with each other and rebounding. The atmospheric pressure can be measured by the height of the column of mercury in an inverted evacuated tube.

One of the first discoveries made with the aid of Torricelli's device was that the pressure of the air is less at high elevations than at low. The pressure at any elevation results from the interplay of two tendencies:

- That of the gas to expand equally in all directions.
- That of each particle to fall toward the earth's surface because of gravity.

The barometric pressure, actually a measure of the weight of the air particles above the barometer, decreases rapidly with height and has only about half its sea-level value at a height of 17,500 feet, thus indicating that only half the mass of the atmosphere is above that elevation, even though the atmosphere extends upward for several hundred miles.

The energy to drive all atmospheric motion is ultimately derived from the sun's radiation. The actual motion is in response to differences in the gravitational attraction of the earth upon air masses of different densities. but the density differences are due to changes in temperature set up by the sun's radiation.

Over a long time the amount of heat received from the sun must be balanced by the heat lost to outer space, or the earth would be either heating up or cooling down. True, climatic fluctuations are not negligible but the fossil record strongly indicates that these are minor; roughly, the heat budget of the earth (amount received minus the amount lost) must be nearly balanced.

Much of the sun's radiation striking the outer atmosphere is reflected back into space; it does not affect the climate or temperature of the earth's surface. Of the radiation entering the atmosphere, part is radiated back into space as heat waves, part is reflected, and a part is absorbed, raising the temperature of the land, air, and water.

Measurements and calculations of heat received at the earth's surface indicate that the amount varies with latitude, much as the ancient Greeks thought. Losses by reflection and back-radiation into space also vary with latitude because a warm body radiates more than a cold one, but these differences are not so great as those in the amounts received. Thus, in low latitudes, more heat is received than lost, and in high latitudes.

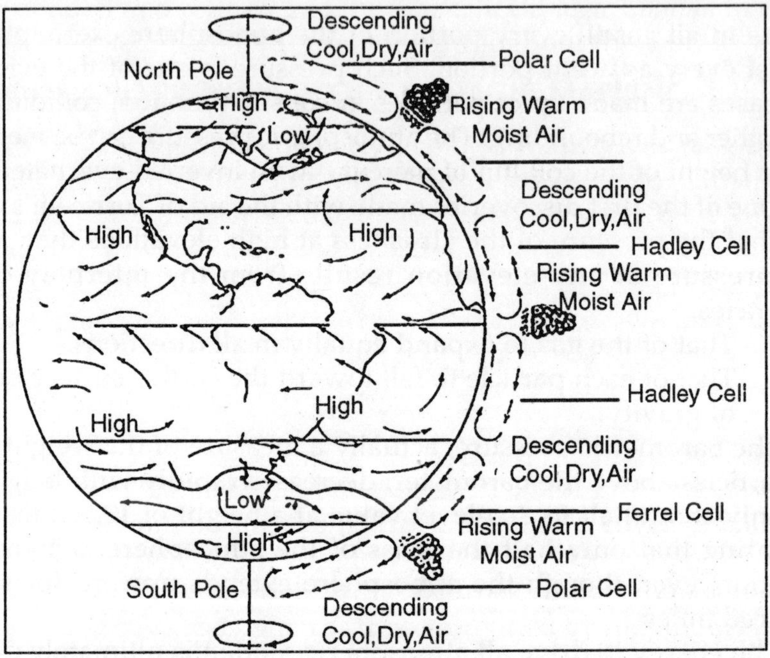

Fig. Atmospheric Circulation in the Northern Hemisphere

Hotter and the polar cooler, the net heat transported from the tropics to higher latitudes must compensate nearly exactly for these differences the last column of the table. Most of this transport is by movements of warmed air (winds) but some is by ocean currents.

Consider first the condition in the tropics: Air heated near the equator must expand (become less dense). A column of air in the tropics thus weighs less than a column of cooler air of equal height in high latitudes. Such a system is unstable, and, since no gas has rigidity, these pressure differences cannot be sustained. The warm equatorial air column rises and flows poleward, while the denser cool air from higher latitudes flows equatorward beneath it. This establishes a convective circulation like that seen in a pot of soup slowly simmering on the stove.

But flow cannot be directly toward the equator because, due to its inertia, the air is deflected by the rotation of the earth.

EFFECT OF THE EARTH'S ROTATION

The equatorial circumference of the earth is about 25,000 miles; a point on the equator thus moves eastward at the rate of more than 1,000 miles per hour. At latitude 40° the eastward velocity is about 750 miles per hour, but at latitude 45° only about 700.

An air mass moving with the earth (on a windless day) has the velocity appropriate to its latitude. If it later moves southward from a point in north latitude toward the equator, its speed is less than that of the ground at the lower latitude, and to an observer there results a northeast wind. In general, if an observer stands with his back to the wind, he finds that in the northern hemisphere all winds are deflected to his right, and those in the southern hemisphere to his left, because of the earth's rotation.

One result of these two factors—the equatorial heating and the earth's rotation—is a prevailing wind called the *trade wind,* blowing obliquely toward the equator in each hemisphere. In the northern hemisphere this is the *northeast trade;* in the southern, the *southeast trade* wind. Between lies the equatorial belt of calms.

If the circulation were a single pair of simple circuits, one in each hemisphere, trade winds would blow all the way from the poles. Actually two (or three) circuits appear in each hemisphere, the upper air coming down in middle latitudes and moving along the earth's surface both toward the equator and toward a pole.

Air Masses

On a smaller scale, air masses, commonly 500 to 1,000 miles across, move slowly in the lower levels of the atmosphere in response to regional or local differences of pressure. Thus, a meteorologist speaks of a cold, dry, polar air mass penetrating beneath a warm, moist, tropical air mass. The advancing margin of such a polar mass is called a "cold front." Rain is likely to occur behind a cold front.

A large polar mass, twisting southeast (dashed arrows) across North America, insinuated a thin advancing wedge beneath a mass of warm moist air moving north and northeast (solid arrows) from the Gulf of Mexico. The lower part of the water-saturated tropic air mass was lifted above the cold heavy mass, just as air rises when a liquid flows into a pan. In rising into a region of lower pressure, the air expanded and cooled.

As cold air can hold less water vapour than warm, the moisture condensed as a torrential rain, bringing devastating floods to the Ohio River Valley and snow to the northern Midwest.

EFFECT OF MOUNTAIN RANGES

Mountain ranges profoundly modify local climates. Moist oceanic winds lose much of their water content as rain or snow when they rise over a mountain range. Such an ascent has the same effect as rise over a cold front.

For example, the windward sides of the Hawaiian Islands may have 50 to 200 inches of rainfall a year; the sheltered sides 5 to 10. Similarly, the deserts of Turkestan are behind the Himalayas, whose seaward slopes receive more rain. The air that pours down the inland slope of a mountain range is compressed and heated, so that precipitation is usually prevented.

CLIMATE TYPES

If one considers rainfall alone, climates may be called *humid* if the mean annual precipitation exceeds 20 inches, and if less than 10 inches, and *semiarid* if between 10 and 20 inches. From another point of view, *maritime climates* are characterized by relatively uniform temperatures, moist air, fog and cloudiness, rainy winter, high wind velocities, and freedom from dust.

In *continental climates*, both daily and seasonal temperature ranges are greater, windless days more frequent, the winter rainfall less, and the air drier and often dust-filled. The extreme continental climate is the and or desert climate, which is much windier than other continental types, partly because of the great daily ranges in temperature. Many deserts lie in "rainshadows" behind high mountains, but some of the greatest—the Sahara, Australian, and southwest African deserts—extend to the ocean shores in the subtropical high-pressure belts.

WEATHERING AND SOILS

The principal significance of the different climatic types to the student of geology lies in their influence on soil formation and on erosion. Each climatic environment places its own stamp upon the soils developed there, and each influences, through its control over vegetation, amount of rainfall, and evaporation losses, the geological processes involved in molding the details of the earth's surface. Erosion, the process of removal of rock waste; here, we will examine the influence of several different climatic environments upon the weathering of different rock types.

The most familiar example of *weathering* is the etching and discoloration of the surface of an unpainted board left out-of-doors. Rock, exposed on the earth's surface, also decays and leaches, but much more slowly. If the product of rock decay is merely broken and discolored, it is called *mantle* rock; but if it is loose and porous enough for plants to find a foothold, it is called *soil*.

Soil is more common than rock at the earth's surface. Almost all outcrops of rock are less firm—more easily crumbled and broken—than is the same rock at a depth of 20 or 100 feet. Many rocks that are black or steel gray where penetrated in mines, wells, or deep quarries are yellow or brown in outcrops. In some, the yellow colour is a mere stain on or near cracks, but in others it is more pervasive and is accompanied by drastic changes in mineral composition of the rock.

That the changes result from weathering is shown by observations on building stones. For example, the exposed faces of the sandstone used in the older buildings at Stanford University turned yellow in 5 to 10 years, and, where exposed to repeated wettings from garden sprinklers, began to crumble in 20 to 30 years.

Analysis of Weathering

Weathering results chiefly from exposure to the air or to the action of substances derived from the air, the most important of which are water, oxygen, and carbon dioxide. Most studies of weathering have been made on one of two very different materials:

- Soils clearly derived from immediately underlying rocks
- Building or monumental stones weathered superficially after years or centuries of exposure under various conditions.

Such studies have shown that weathering usually includes both *mechanical disintegration* and *chemical decomposition. Disintegration* includes only loss of coherence, with little or no changes in composition. It does not include abrasion and removal of the constituents; such movements are part of erosion.

Decomposition or *chemical weathering* is the term for change in chemical (and mineralogical) composition. The complex silicates that make up the bulk of most crystalline rocks alter into other substances, such as hydrous silicates, hydrous oxides, and carbonates.

EXAMPLES OF MECHANICAL WEATHERING

Frost Action

Wherever water freezes, the 9 per cent expansion is an effective disruptive force. The expanding of ice to break rock is called *frost action* or *wedging*. Water freezes in the soil and in cracks or pores of rocks. The colder the climate, the deeper the freeze.

Thawing and repeated freezing add to the effect, however, so that frost action is more effective in cold temperate zones than under extreme arctic conditions.

In soil, especially, because of the high water content and the complexity of the pore systems, freezing increases the volume considerably. After thawing, the soil is open and spongy. To break rocks, freezing water must be confined. As the water in a crack begins to freeze at or near the ground surface, it may be completely confined in the lower part of the crevice in the rock. Over a long time, rocks become extensively shattered by frost action. Many high mountain peaks are covered with rubble formed in this way.

Plants and Animals as Aids to Weathering

Organisms assist in the breakdown of rock to soil. Their action is in part disintegration, and in part chemical effects. The roots of growing plants powerfully wedge the soil aside, raising and breaking slabs of rock (or concrete), and widening cracks. Burrowing animals move and mix the soil effectively.

Charles Darwin, in his last published work, showed that English earthworms may spread their casts over the ground to the average thickness of 0.1-0.2 inch per year. The soil is thus loosened, aerated, and subjected to the chemical action of the worms' digestive processes. This mixing probably accounts for the uniformity of the humid, and especially the forest or meadow, soils. There are practically no earthworms in arid regions.

Most plant tissues are carbohydrates (compounds of carbon, hydrogen, and oxygen). Microscopic organisms, such as molds and bacteria, in the absence of oxygen, change leaves, fruit, and wood on or in the soil into the dark organic substance known as *humus*. In the presence of oxygen, the organic substances are destroyed and CO_2 and H_2O are formed from them. Some H_2CO_3 is formed by both processes and contributes to chemical weathering. Living or dead, land plants and animals are aids to weathering.

Other Agents of Disintegration

Frost action and wedging by plants are the most effective agents of disintegration, but many other processes play a role.

Chemical decomposition is one cause of disintegration, and possibly a major one. Feldspars swell as they weather into clay, and many other minerals decompose to substances that occupy more volume. In a granite composed of feldspar, quartz, and mica, this swelling of the rotting feldspar may disintegrate the massive rock to a pile of granitic sand. Alternate swelling and shrinking of the clay as it is wet by the rains and heated by the sun aid the process.

Rain may also wash soluble salts into cracks in rocks and soils where they crystallize. The growing crystals may exert enough force to break and disintegrate the rock still further. Telephone poles set in the Bonneville salt flats west of Salt Lake City show this effect to a striking degree. During the wet season the pole acts as a wick, drawing the salt brine up into the wood by capillarity. Here, the brine evaporates, precipitating salt crystals so abundantly that, after a few years, the pole up to about 18 inches above ground is swelled to twice its normal size, and the wood fibers are completely shattered.

A forest fire often shatters and spalls the edges from exposed rock ledges and boulders. Rock is a poor conductor of heat, so the interior is not appreciably warmed by the fire. Consequently, the highly heated outside expands and pulls away from it. Lightning also breaks and shatters rocks.

It is possible that the drastic day-to-night changes of temperature that occur in deserts may cause enough expansion and contraction to weaken or even break some minerals and rocks.

However, specimens of granite have been placed in electric ovens and run through the range of temperatures characteristic of deserts many thousands of times without showing any appreciable loss in strength. Doubtless, the swelling effects of decomposing minerals and the expansion of growing salt crystals are far more important in producing disintegration in deserts than are the diurnal temperature changes.

EXAMPLES OF CHEMICAL WEATHERING

We will now consider a few details of the chemical processes whereby rockforming minerals are decomposed into hydrous silicates, oxides, and carbonates. Our first example is the chemically simple rock, limestone. Weathering of Limestone. Limestone is almost wholly calcite, but generally contains a few particles of clay. Calcite dissolves very slightly in pure water, sending some Ca^{2+} and CO_3^{2-} ions into solution. As indicated in chemical notation:

$$CaCO_3\ CO^{2+} + CO_3^{2-}$$

In water containing dissolved carbon dioxide (as all rain water does), the calcite becomes much more soluble. The water and carbon dioxide react to yield carbonic acid and this in turn furnishes ions of hydrogen and bicarbonate. Thus:

$$H_2O + CO_2\ H_2CO_3\ H^+ + HCO_3^-$$

But the hydrogen ion thus formed reacts with carbonate ion from the calcite to form more bicarbonate ion:

$$H^+ + CO_3^{2-}\ HCO_3^-$$

The arrows in these equations are the chemist's shorthand for a reversible process. In solutions the combination and dissociation of ions are constantly proceeding at rates governed by the abundance of the several ionic species present.

For example, at equilibrium in reaction as much calcite is being precipitated from solution as is being dissociated into ions in that solution. But if the ions on one side of the reaction are selectively removed, they obviously cannot furnish as many products on the other side as they formerly did—the reaction will take place in the direction of yielding more of the ions being removed.

When calcite is dissolved in water containing carbon dioxide, the carbon dioxide enters into new combinations with the water, as in equations, so that more of the products on the right side of equation are formed than would be in pure water; that is, more calcite is dissolved. Thus, in the weathering of limestone the most important reaction is the slow solution of the calcite in percolating water that has absorbed carbon dioxide from the air.

Calcite is dissolved not only at the surface, but along every crack and fissure into which the solutions penetrate. The clay in the limestone accumulates on the surface as the underlying rock slowly dissolves. A few inches of clay soil may thus represent the residue of tens or scores of feet of dissolved limestone. Where underground circulation is vigorous the limestone may be extensively dissolved, resulting in the formation of great subterranean caves.

It is noteworthy that, despite the differences in composition of the underlying rock, soils derived from limestone and granodiorite are both rich in clay. The clay minerals are among the stablest under surficial conditions, both in temperate and in arctic regions.

From this analysis of the processes of weathering we can see that disintegration and decomposition take place together, each aiding the other. In moist temperate regions, which are the areas that have been most studied by soil scientists, decomposition appears to be much the more important. Disintegration by frost action, however, is widespread, and in cold regions it may predominate.

EXFOLIATION

The splitting away of successive scalelike layers of rock from an exposed or soil-covered surface of massive rock (not a platy rock such as shale or schist) is called *exfoliation*. The separated sheets or plates may be flat or curved, paper thin or many feet thick, a fraction of an inch or hundreds of feet long. Two types of exfoliation may be distinguished—small thin flakes on massive rocks of all kinds; and giant plates, usually on granite or granodiorite.

Closely fitted blocks of dolerite, sandstone, granite, and other mediumor fine-grained rocks frequently develop numerous thin concentric layers of partially weathered material. During weathering, the material expands as hydrated clay minerals are formed from the original minerals. Probably the outermost shell weathered and expanded first, pulling away from the fresher rock beneath. The shells formed successively inward in similar fashion. In each shell, chemical weathering preceded the mechanical process of exfoliation.

The giant curved plates of fresh rock that have split away in forming the Yosemite domes in California, Stone Mountain in Georgia, and Sugar Loaf at Rio de Janeiro, indicate that exfoliation may take place without much weathering. These domes are all in plutonic rocks that originally consolidated beneath a thick cover of overlying rocks. Perhaps unloading by erosion of thousands of feet of the cover has permitted upward expansion and the formation of cracks parallel to the surface.

Colors of Soils

Most soil colors are due to iron minerals or organic matter. Iron exists in two states: ferrous and ferric. A ferric ion may be formed from a ferrous ion by the losing of one electron, a process called *oxidation*. When hydrated ferric oxide ("limonite") forms by chemical decomposition of a ferrous silicate such as biotite, the reaction is made possible by the addition of oxygen from the air. The finely divided "limonite" stains the resulting soil yellow. If the soil is repeatedly dried out, as in a climate with a warm dry season, the "limonite" may lose its water and change to red hematite; the soil then becomes red.

Iron is reduced to the ferrous state as easily as it is oxidized. A pale greenish or a dark gray soil usually indicates that the iron has been reduced by plant residues or other reactions with organic matter. Ordinarily this occurs where the air is excluded because pores have been filled with water. Thus, a black soil may indicate swamp conditions; the iron has been reduced and the soil is rich in humus. Colors of soils give clues to the conditions of weathering and to the abundance of organic matter.

RESIDUAL SOIL AND THE SOIL PROFILE

The surface changes in a fairly uniform rock brought about by weathering are well illustrated in the Sierra Nevada of California. The canyon walls expose hard gray granodiorite, composed chiefly of plagioclase, with smaller amounts of quartz, orthoclase, biotite, and hornblende. The rolling forestcovered uplands between the canyons have a red-brown soil. The surface layer, called by soil scientists the

"A horizon," is a sandy loam (a mixture of quartz sand, silt, minute clay particles, and decomposed plant residues).

In the fine-grained reddish matrix are embedded many irregular quartz grains, alike in size and shape to those in the granodiorite. At about a foot depth, the clay content increases. This sandy clay is the subsoil or "B horizon." At two to four feet, the soil becomes paler and sandier. The sand grains include not only quartz, but abundant feldspar, stained but readily recognized by its flashing cleavage faces, and also many micaceous flakes.

These are not jet black, as is the mica of the granodiorite, but iron-leached pearly yellow scales sometimes mistaken for gold. Still deeper, this material grades imperceptibly into stained and crumbly material in which the texture and characteristic minerals of the granodiorite can be vaguely recognized. The transitional material below the zone of high clay content (B horizon) is called the "C horizon.".

The conclusion seems inescapable that the surface material is developed by weathering of the underlying granodiorite. Such a soil, still in place on its source rock, is called *residual*. The A horizon of the residual Sierra soil is looser and sandier than the compact B horizon. Apparently some clay formed by decomposition of feldspar near the surface has been washed downward from the A horizon, giving the B horizon an extra portion.

In contrast with the residual soil of the Sierra granodiorite is that of certain parts of Kentucky where the bedrock is limestone instead of granodiorite, and the climate, though also temperate, is more humid. The A horizon consists of black humus-rich clay; the B horizon of light-gray limy clay which passes downward into white limestone.

Most soils show characteristic "profiles" composed of two or more distinct layers or soil horizons, of which many kinds have been distinguished from various climatic environments by soil scientists. Comparison of various soils is facilitated by comparison of their respective soil horizons.

SOILS DEVELOPED ON TRANSPORTED MATERIAL

Many soils have developed not on bedrock, but on unconsolidated material deposited by streams, glaciers, or other agents of erosion. Such materials have generally been at least partially weathered before deposition. After deposition the loose material allows easy entry of air and water, and so weathers quickly to form typical A and B soil horizons.

The soils of the Great Valley of California afford an example. The residual soil of the Sierra Nevada, developed from granodiorite, has been washed into streams and transported to the Great Valley. On debouching from the mountains, some streams have deposited hundreds of feet of soil and rock debris at the mountain base. In places, such sediments, even some gravels, have been left undisturbed long enough to have rotted into brown soils with a characteristic soil profile.

The brown silty A horizon, two or three feet thick, contains scattered quartz pebbles and numerous groups of closely spaced angular quartz grains that are probably relics from decayed granodiorite pebbles. The compact B horizon, one or two feet thick, has been formed in part by the filtering down of minute clay grains from the A horizon. Such compact clayey B horizons, especially when cemented by iron oxide, calcium carbonate, or other bonding materials, are called *hardpan*. The C horizon consists of the relatively unaltered gravel or other parent material below the B horizon.

Another example of soil developed on transported material is that in eastern Massachusetts derived from rock flour (very finely ground fresh rock), boulders, and clay transported by glaciers in the geologic past. Here, the A horizon consists of three or four inches of dark humus-rich soil overlying sterile rock flour and fragments.

CLIMATIC FACTORS IN WEATHERING

In Moist Temperate Climates

The weathering of limestone and granodiorite produces mainly minute grains of one or more of the clay minerals, with more or less quartz sand, iron oxide, flakes of altered mica, and perhaps some partially decomposed feldspar. For this climate, a rough, but not precise, rule is that the more iron and the less silica in a rock, the more rapid its weathering. Thus, basalt weathers readily, but quartzite is almost immune. Limestone weathers rapidly in this climate.

In Dry and Cold Climates

From the relatively few investigations reported it appears that disintegration is the chief weathering process in the driest deserts, such as those of southwest Africa. It is also dominant in polar latitudes and on the high mountain peaks of the temperate and torrid zones. In arctic regions, frost action predominates; in the subtropical deserts, shattering of unknown origin prevails, perhaps caused chiefly by slight chemical changes, facilitated by expansion of the surface layer by decomposition, and by solar heating. Obvious chemical changes are minor and fine-grained soils rare. Vegetation is sparse; hence, carbonic and other soil acids are of less than usual importance. In subarctic regions such as

Finland, however, chemical weathering of a special type, producing very siliceous soils, appears to predominate, especially where vegetation is fairly abundant and its decomposition slow and steady.

Chemical weathering of another sort, producing soils rich in calcium carbonate and high-silica clay minerals, is predominant in semi-arid regions such as the western interior of the United States. The calcium carbonate accumulates because, during most of the year, the soil water rises to the surface through small connected (capillary) pores, and releases its dissolved carbonate by evaporation, rather than draining away through subterranean pores, cracks, and channels.

A striking contrast between moist and semi-arid weathering is furnished by limestone, which dissolves to leave pock-marked valleys in humid regions but stands out as bold ridges, almost like quartzite, in a semi-arid environment like that prevailing over most of Arizona and Nevada.

In the Tropics

The rain forests of the Congo and Amazon basins grow in soils which have not been adequately studied, but which appear similar to the aluminum-silicate clay soils of moist temperate regions.

The grass- and tree-covered savanna lands characterizing wide belts north and south of the tropical rain forests are largely underlaid by yellow and red-brown soils called *laterites*. The characteristic dark-brown upper portions of laterites dry to bricklike hardness and are sometimes used as building materials. Typical laterite areas are found in India, Nigeria, Central America, and Brazil.

Because laterites offer very different problems than do the products of weathering in temperate climates, and because they illustrate one way in which a chemical element such as iron or aluminum may be concentrated into valuable ores. Laterites vary considerably in mineral composition, but are typically aluminum and iron hydroxides and oxides mixed with some residual quartz. A rare variety called *bauxite* is almost pure $Al_2O_3 \cdot nH_2O$ and, hence, valuable as the ore of aluminum. In laterites, practically all the silicon of the original silicates has been leached out by rain waters, along with the easily soluble sodium and potassium, and the acid-soluble calcium and magnesium.

The chemical aspects of this are difficult to understand. Silica is normally soluble only in water containing many more OH^- (hydroxyl) ions than hydrogen ions (an alkaline solution). However, calcium and magnesium are not soluble in such a solution, but in those with excess hydrogen over hydroxyl ions (acid solutions). Yet all three of these elements have clearly been leached from the soils.

Most laterite regions have marked wet and dry seasons. It has been suggested that during the warm dry seasons organic acids are so completely oxidized (to carbon dioxide and water) that the dry soil contains no acidproducing materials because the carbonic acid goes into the atmosphere as a gas. But the soil may retain a very small amount of material which would, when dissolved, give an alkaline solution. Then, at the onset of the first rains, silica is carried off in the temporarily alkaline soil solution before the decay of new vegetation again restores the carbonic acid supply.

Residual laterite soils are characterized by a pale *zone of leaching* just above the parent rock, and a dark-brown *concretionary zone* at or near the surface. Each zone is usually a few feet or a few tens of feet thick, locally deepening to hundreds of feet. The concretionary zone is a concretelike mass, composed chiefly of either dark-brown "limonite" or of numerous limonitic nodules (*concretions*) of pea or marble size, more or less well cemented into a solid mass. Possibly the concretionary zone should be called the "B_1 horizon," and the zone of leaching the "B_2." The uppermost parts of some lateritic soils are crumbly or even sandy (because of unaltered quartz) and perhaps represent the A horizon.

In laterites some elements needed by plants have been leached away, and others, such as phosphorus and iron, are precipitated as iron phosphates or other compounds so insoluble as to be practically unavailable to plants. Therefore, pineapple plants grown on the lateritic soil of the Hawaiian Islands are sprayed with soluble iron salts, and small quantities of soluble phosphates are applied frequently.

What happens to the silica leached out during lateritization? No definite answer can yet be given because the subtropical distribution of even so abundant a substance as silica is little known. Extensive siliceous crusts have been reported from Angola, just south of the Congo Basin, and from other parts of tropical and subtropical Africa, but laterite is not definitely known near them.

When the total amount of silica in certain rocks is compared to the amount in the streams that drain them, it appears probable that silica is more rapidly removed in tropical climates. It has been shown, for instance, that the relative solubility of the silica in certain igneous rocks in tropical British Guiana is about twice as great as the world average for all rocks in all climates.

RATES OF WEATHERING

The rate of weathering varies greatly with different rocks and different climates. In less than eighty years the Edinburgh inscription on marble in memory of Joseph Black, the discoverer of carbon dioxide, was rendered illegible by the action of carbon dioxide and other acid-forming substances. On the average, about nine millimeters of rock per hundred years has dissolved from faced limestone in Edinburgh.

Furthermore, there has been marked disintegration of the limestone by frost action. In small Scotch towns, with less coal smoke, the marble in tombstones has been dissolved more slowly, but the effects are notably greater on limestone than the barely perceptible roughening of slate slabs in one or two hundred years.

Europe, southeast Asia, and Central America are dotted with ruined stone castles or temples. For many of these structures the dates of building and bandonment are known quite accurately. A few ruins have been studied by soil scientists or archeologists. Kamenetz fortress in southern Russia was built in 1362 and remained in use until 1699. Thereafter, through neglect, the stone blocks began to disintegrate and decompose. The top of one limestone tower developed a soil 4 to 16 inches thick in the 231 years between 1699 and 1930. This soil was much like that on undisturbed, similar limestone in the vicinity.

Other towers and castles under the European climatic environment have weathered more slowly, especially if constructed of sandstone, granite, or other high-silica rocks. Most observers consider that the rate of chemical weathering is probably most rapid in moist tropical climates, even though the evidence from dated structures is inconclusive.

The arched roofs of Angkor in Cambodia, Indo-China, stand intact after seven centuries of neglect. Though the jungle crowds close, and clumps of green plants grow in the crannies, the endless vistas of sculptured walls indicate only slight weathering of the rock. In a somewhat less luxuriant coastal jungle southeast of Vera Cruz, Mexico, a slab of basalt, definitely dated by archeologists, lay in moist soil for at least 400 years without blurring of the Mayan inscription, whereas other inscriptions at the same place, with an almost identical history, have become illegible.

Weathering proceeds most slowly in a dry warm climate. Inscriptions carved between 3 B.C. and 79 A.D. above the doors of sepulchers in the crumbly sandstone cliffs of northwest Arabia could still be read in 1876. Forty centuries ago, not far from the site of Assuan Dam in Upper Egypt, a surface of Syene red granite was smoothed by men's tools and dated by an inscription. This surface, though exposed to direct sunlight, is still firm.

At various dates, ranging from 2850 to 313 B.C., colossal statues from this quarry were set up at Luxor and elsewhere in the somewhat moister atmosphere of Middle and Lower Egypt, and blocks of the same rock were used to face pyramids near Cairo. The average rate of exfoliation for all the Syene cut granite studied by D. C. Barton, an American visitor to Middle and Lower Egypt in 1916, was about 0.5 to 1.0 centimeter per 5,000 years. In general, the older monuments showed the greater effects, but the maximum measured exfoliation was 1.0 centimeter in the youngest blocks studied (313 B.C.).

Two obelisks of the Syene granite, each bearing many deep-cut hieroglyphics, and each now called Cleopatra's Needle, stood about 3,500 years at Heliopolis and Alexandria in Egypt with only slight weathering. One, removed to London, has weathered appreciably but not disastrously. The other needle, brought to New York about 1880 and set up in Central Park where it is exposed to frost as well as frequent wetting, exfoliated so extensively that by 1950 (despite the application of shellaclike preservatives in the nineteen-twenties and thirties) part of the pictured story was completely erased. More weathering had taken place in 70 New York winters than in 50 times 70 Egyptian years.

Chapter 3

Rocks on the Earth Surface

A rock may be defined as a mineral mass or an association of minerals, which in their natural occurrence form an essential part of the earth's crust. This distinction is not a very precise one, especially when the material of the rock consists of only one mineral.

Thus the calcite in a vein would be considered a mineral, while essentially the same material in a bed of marble would be considered a rock. We proceed with the rocks, that in practice the distinction between rock and mineral can as a rule be readily made.

AGE RELATIONS OF ROCKS

It will be useful at this early stage of our study to recognize the fact that the rocks of the earth's crust are of various ages. Some, like the rocks which make up the Adirondack Mountains, Pikes Peak in Colorado, the Highlands of the Hudson, the Scottish Highlands, the main mass of Finland, and a great part of central France, etc., are very old; others, like those of the "puys" which are scattered over the central French region, the basalts of the Columbia and Snake River plateaus of the northwestern United States, the rocks immediately underlying Paris, London, Vienna, and Berlin, and the rocks of southern Florida are very young, though not all of the same age.

It is possible to divide the history of the earth into a number of periods and eras, just as human history can be divided. But whereas the successive periods of human history are measured by centuries at the most, those of the pre-human earth history are measured by hundreds of thousands if not by millions of years. And as we can refer the monuments and buildings of human origin to their successive periods in human history, often from the character of these monuments and buildings themselves, so we can refer most of the different rocks and rock structures of the earth's crust to their respective geological periods, or to the period of the earth's history when they came into existence as rocks. It is desirable that the student should begin to familiarize himself at this point with the names of the different periods of earth history At a later stage of his studies he will learn by what means it becomes possible to refer rocks and rock structures to their proper period. When in the succeeding pages we refer to the geological age of any rock mass the student should consult the table until he has become familiar with the succession of the periods.

BED-ROCK AND MANTLE-ROCK

In general we may distinguish between the solid or bed-rock and the unconsolidated rock or rock material, which latter is commonly called the mantle-rock because it covers or mantles the bed-rock which everywhere underlies it and projects through it as ledges. The mantle-rock is of course much younger than the bed-rock upon which it rests. In the northern United States and Canada and the northwestern part of Europe, the mantle rock generally rests abruptly upon the solid or bed-rock, the contact line between the two being commonly sharp.

In many other por tions of the earth, however, there is a gradation between the two, the mantle-rock becoming more stony downwards, and passing into rotten rock and finally into fresh bed-rock. This present one will only take account of the solid rock which forms all except the surface film of the crust of the earth.

CLASSIFICATIONS OF ROCKS

General Principles of Classification

Classification of natural objects may be made either on a natural or on an artificial basis. A natural classification is based on the origin and development of the objects classified or on their *genesis*. Such a classification is therefore called a *genetic* one, and it alone has a permanent scientific value. It is true that a genetic classification can be made only when the genesis or origin of the objects classified has been determined, and that therefore the artificial systems of classification will always precede the scientific or genetic one.

The artificial classification is generally based upon the possession in common, by the objects classified, of a single character or, at most, a few characters. Thus whales were formerly classified with fish because, like these, they lived in the water and had a fish-like form. Their true relationship is, however, more closely with elephants, both being mammals. The *eel-grass* of our coast is not a grass, but belongs to the family of water lilies, although its leaves are grasslike. Rock-salt and sandstone have little in common, though generally classified together as sedimentary rocks; both may have been deposited in lagoons near the seashore, and therefore in a sense are sediments.

In general, it may be said that the progress of any science is indicated by the replacement of the original artificial by the natural or genetic classification. Thus as the study of plants developed into the science of botany, the original artificial classification of plants, based on the number of stamens of the flower, and on other superficial characters, became superseded by the natural classification, which is based upon community of origin of the members of the same group; and in zoölogy, we find that artificial classifications, based on superficial resemblances or the possession of a common character, are constantly discarded as the true *genetic* or natural relationships of animals are becoming more fully understood.

Artificial Classification of Rocks

A convenient classification of rocks, which has come into general use, recognizes three fundamental divisions, as follows:

Igneous Rocks

Rocks which result from the cooling of a molten mass, or *magma*, either upon the surface of the earth, or within the crust at greater or less depths. An example of the first is *basaltic lava*; of the second, *granite*.

Sedimentary Rocks

Rocks which were formed:

- As mechanical sediments in water or air.
- As chemical precipitates and evaporation products from solution in water.
- As deposits formed by organisms either in water or air.

Examples of these are:

- Sandstone, shales, etc..
- Rock-salt, gypsum, cave deposits (stalactites), etc..
- Coral and shell limestones, chalk, guano beds, coal.

The three groups here indicated are commonly made subdivisions of the sedimentary rocks, and designated respectively as follows:

- Mechanical sediments.
- Chemical sediments.
- Organic sediments.

Metamorphic Rocks

These rocks were originally members of one or the other of the preceeding divisions, but have become sufficiently altered by natural agencies, such as pressure, heat, and so forth, so that for the most part the original characters are obliterated or lost, and new and special characters are added. Examples of such metamorphic rocks are:

- *Gneiss* – which may be derived from an original granite, but may also have originated from another rock, even from a sediment.
- *Mica schist,* which may have been derived from a shale or an impure sandstone or some other rock.
- *Marble,* which is derived from some form of limestone that may have been of organic, of chemical, or of mechanical origin.
- *Graphite,* which may be derived from coal beds or from carbonaceous shales, etc.

Such a classification is sufficiently serviceable for all practical purposes, but it cannot be called a scientific classification; because in the second and third group are included types of very diverse origin; for the three groups mentioned under sedimentary rocks have little or nothing in common, except that they are sediments in a very broad and indefinite comprehension of that term.

More correctly speaking, they are not igneous rocks, and not pronouncedly metamorphic rocks, and this is almost their only claim to a grouping under a separate division. The metamorphic group, too, includes, as we have seen, rocks of very diverse origin, but since it is not as a rule possible to determine, except perhaps after prolonged and careful study and analysis, if then, what the character of the original rock was, it will probably always be necessary to retain this group as a matter of convenience.

GENETIC CLASSIFICATION OF ROCKS

While the practical worker may find the preceding classification sufficient for all needs, and while in the succeeding we may frequently refer to these convenient three types, the student should nevertheless understand the principles of a natural classification, and so far as it is practicable we shall base upon such a natural classification. For in geology as in other sciences, logical thinking is of the first importance, and logical thought is best fostered by the most rigid adherence to exact methods of classification, and the only exact classification of natural objects is one based upon genetic relationship, that is, upon community of origin of the members of each group.

At the outset of our endeavor to understand the natural relations of rocks we must clearly comprehend two fundamental principles. The first of these is, that in this natural world of ours all things are subject to continued change, even though that change may be a very slow one, so slow that the years of a man's life or those of many successive generations are of insufficient length to permit the recognition of such changes. All rocks are constantly undergoing modification, and though sometimes these changes may be rapid, as in the coking of coal in a burning mine, or the change of clay to brick in a kiln, the more usual method of change is a slow and gradual one. In a certain sense all rocks may be considered as *metamorphic* or altered rocks, a view strongly held by some geologists.

Extreme metamorphism, such as produced the common types which are usually spoken of as metamorphic rocks, is only a phase of the general alteration or metamorphism which all rocks undergo, and this phase is characterized by the greater or complete obliteration of the characters which the rock possessed during its early history. This more pronounced change may have been brought about by the greater intensity of the activities responsible for it, or by their longer continuance in time, or by both.

The second principle, a knowledge of which is fundamental to our understanding of rocks as well as all other natural objects, is that related types are not separated as a rule by sharp lines of demarcation, but that always and everywhere in nature gradation is the predominant rule.

Thus a sandstone may grade into a shale, on the one hand, and into a limestone on the other, and shale and limestone may intergrade, though of course there are always types which retain the characters of one or the other with sufficient distinctness to make their classification possible.

A granite may grade into a syenite, and this into a diorite; and again, a granite and a diorite may approach each other so closely in character that classification becomes difficult. In like manner a sandstone may have so many characters of a schist that it becomes a difficult question to which of the two divisions it belongs. A shale may grade into a slate and a granite into a gneiss. With slightly metamorphosed rocks, classification is often very difficult, though a thoroughly metamorphosed rock is easily recognized as such.

Extreme metamorphism is only a later or final phase of the change which all rocks are undergoing, and in a natural classification such end-products are to be placed with the group from which they are derived. Two gneisses, for example, one of which is derived from a granite and one from some type of sandstone, are not related to each other any more than two human beings are related because they speak the same language, or obey the same laws of civilization.

As before said, however, the practical difficulty of determining the original character of a metamorphic rock must be reckoned with, just as in the study of thoroughly civilized human beings of different but entirely disguised nationalities, the practical difficulty of ascertaining the nationality of each (assuming that the individuals refuse or are unable to disclose it) must be taken into account. In this respect thoroughly metamorphosed rocks hold the same relation to their ancestral type, as thoroughly Americanized individuals descended from different nationalities hold to their ancestors.

It thus becomes necessary for us to study first the unaltered or but slightly altered rocks, the history of which is ascertainable from the characters which they retain.

THE UNALTERED OR LITTLE ALTERED ROCKS

When we study the unaltered or slightly altered rocks, we again note that they fall into two quite readily recognizable divisions. The first of these comprises rocks which have been produced from material not originally rock, — for example, by cooling from a lava, crystallizing from a solution in water, and the like, — a group which in a broad sense may be spoken of as produced in a chemical manner.

The other division comprises rocks which are made up of fragments or particles of other rocks, such as sandstone made of grains of sand, a conglomerate made of pebbles of various kinds and of sand, and others like these. This is the group of fragmental or *clastic rocks* which are always made from some preëxisting rock that has been broken down into discrete particles, which then are recombined either directly or after more or less assorting.

The first group, on the other hand, is that of the non-fragmental or non-clastic type, which is not formed from fragments of other rocks, but from non-rock material. In the regular order of formation this type would appear first, though it is perfectly possible that non-fragmental rocks may originate from material which itself was derived from other rock, by melting, solution, or vaporization of such a rock and subsequent resolidification. But in this case there is an intermediate non-rock stage, while the material of the fragmental or clastic rocks is always rock, though it may be broken into the finest particles. We shall consider the characteristics of each group briefly.

The Non-Fragmental Rocks

The non-fragmental rocks may be properly regarded as contributions to the lithosphere from three of the other spheres; namely, the hydrosphere, the pyrosphere, and the atmosphere. This contribution may be direct, as in the case of the hardening of a lava from the pyrosphere, the separation out of salt from the water of the hydrosphere, by the concentration of that water or by chemical reactions, or the separation of snow or hail-stones by the solidification of the water vapour from the atmosphere.

Each of these spheres may thus in a measure be credited with the generation of these respective rocks, and it is therefore possible to speak of these three types, respectively, as: *pyrogenic,* generated by the pyrosphere; *hydrogenic,* generated by the hydrosphere; and *atmogenic,* or generated by the atmosphere. Instead of these names we may of course speak of these deposits as of igneous, aqueous, and atmospheric origin, respectively, recognizing the fact that the rocks contributed by the atmosphere direct, that is, the snow and hail, are of an evanescent character. In addition to these three types, — the direct contributions of three of the inorganic spheres to the lithosphere, — there is an indirect contribution of material from two of these spheres by the agency of organisms. Both animals and plants take lime from the sea-water to build their shells and other calcareous structures, and other organisms take silica from sea-water to build their hard parts. Deposits of such organically secreted lime and silica are very important in the construction of the lithosphere, and form, as a rule, readily recognizable types of rocks.

Again, carbon is taken by plants from the carbon dioxide of the atmosphere, and the plant tissues built up from it may often become compacted into beds of coal which form important members of the rock series of the lithosphere. Here again is an organically formed rock, the material of which is obtained from the atmosphere.Such organically formed rocks are the contribution of the biosphere to the lithosphere, and in conformity with the method of designation which implies such a relation to that sphere, they may be called *biogenic* rocks.

We may then summarize in the following table these four, fundamental types of non-elastic rocks, rocks which are produced essentially by chemical (including physiological) reactions from nonrock material. We term these rocks *endogenetic* because they are formed by forces in a measure inherent in the material of the spheres which produce them.

TYPES OF ENDOGENETIC ROCKS

- Pyrogenic or Igneous: Produced by the pyrosphere.
- Hydrogenic or Aqueous: Produced by the hydrosphere.
- Atmogenic or Atmospheric: Produced direct by the atmosphere.

- Biogenic or Organic: Produced by the biosphere from material taken from the hydrosphere or atmosphere.

We may consider these rock types from another point of view. Were we to imagine the rock material of the earth deprived of its solid character, that which is essential to the constitution of a rock, we should have to think of it as existing in one or more of three possible forms.

It may be turned into a molten mass like lava, by the application of heat, and so constitute an igneous magma, in which all the minerals which go to the making of the rock would become as it were dissolved in one another, and the compounds dissociated into their ions. It would then become a part of the pyrosphere, and in this state most of the rocks of the earth are believed to have existed at an earlier period according to the believers in one hypothesis of the earth's origin when the earth as a whole was a molten mass. Whether this theory is correct or not, the fact that many of the rocks of the earth's crust were at one time in this state cannot be doubted. By solidification *igneous* or *pyrogenic* rocks are formed.

The rocks of the earth may be turned into a condition of vapour — by the application in most cases of still greater heat. In such a state they would become a part of the earth's atmosphere. The same hypothesis of earth-origin holds that this was a condition of the entire rock mass of the earth at a still earlier period, and that from this condition of vapour was separated at a later period the molten material of the earth, and later still the water of the hydrosphere.

Again, the truth of this hypothesis is not essential to the recognition of the fact that some at least of the rocks of the earth were formerly in a condition of vapour, and that they were separated from the earth's atmosphere either by direct condensation to-day in the formation of snow, or by the work of organisms, such as has resulted in the separation of the carbon of the atmosphere, which now constitutes our coal beds, by the agency of plants. Rocks derived by direct condensation from the atmosphere are *atmogenic* rocks, while those separated by the activities (physiological) of organisms, are *biogenic* or organic rocks.

Finally, we may think of the rocks of the earth, or at least some of them, as dissolved in the universal fluid envelope of the earth — the water — and so become a part of the hydrosphere. If rock material, thus held in solution, is separated out from the water by direct condensation, by chemical reactions, or by electrolytic action, *aqueous* or *hydrogenic* rocks are produced. If, however, the dissolved material is separated out by the agency of organisms, as in the formation of limestone masses on coral reefs by polyps and lime-secreting seaweeds (algæ), it becomes an *organic* or *biogenic* rock.

There are no other primary states than those of fusion, solution in water, or vaporization, into which the rocks of the earth may be changed, nor are there any other known ways in which rocks are formed from the three states of primary dissociation except by direct precipitation or separation or by organic agencies. Hence the four rock types — the igneous or pyrogenic, the aqueous or hydrogenic, the atmogenic, and the organic or biogenic — are the only primary types recognized.

The Fragmental or Elastic Rocks (Exogenetic Rocks)

These are the rocks which are made up of fragments of other rocks, which may range in size from dust particles of microscopic dimensions to boulders many feet in diameter. For their production it is evident that preëxisting rocks should be broken into fragments, and that these fragments should become recemented or bound together again in some manner.

The methods by which rocks are broken into fragments, and those by which the fragments are recemented will be taken up later, but we may here use as an illustration of this type of rock one of the artificial rock-making processes carried on by man, and which differs primarily from the natural process in the rapidity with which it is carried forward. This is the process of manufacture of rubble concrete for paving and other construction.

Rocks like the trap of the Palisades, an ancient igneous rock, are quarried and passed through the stonecrusher, where they are reduced to rock rubble or particles of small dimensions. They are assorted into different sizes by screening, and bound together by a cement, which is a mixture of lime, alumina, and silica, and of sand, which is the finer rock material obtained either from the screening, or more commonly from natural sand banks.

The result is a elastic rock, but an artificial one, and this is produced in a few days, whereas a similar rock produced in nature might require as many centuries or millenniums for its production. But trap rocks or other igneous rocks are not the only ones used in the making of rubble concrete, nor are igneous rocks the only source of elastic material in nature, though they are often the most common one. .

All rocks, — igneous, aqueous, and organic, and elastic rocks as well, — are broken into fragments by natural agencies, and from these fragments new elastic rocks are made. For the rocks of the earth's crust are being constantly reworked, all of them, whether elastic or non-elastic, furnishing material for the formation of a younger bed of elastic rock. Let us take an example for illustration from the eastern United States. There is in the foothills of the Catskill Mountains a great deposit of bedded elastic rocks known to the arts as Hudson River Blue Stone, and used for the manufacture of flagstones for sidewalks in New York City and elsewhere, for curbings, and for many other purposes.

Examination under the microscope shows that this rock is composed of small particles, some of which are themselves small fragments of elastic rocks (these are called *clastoliths*) and this shows that this particular blue-stone rock is made up of material derived in large part, if not wholly, from an older elastic rock which was broken into fragments, assorted according to size by natural agencies, and recemented to form new elastic rocks, the "Blue Stone" being made up of the assorted finer particles only.

From the character of the small fragments of elastic rock (the clastoliths), and from a study of the structural and age relations of the "Blue Stone" to the rocks of greater age, it has been possible to determine that the material of the "Blue Stone" was derived from the so-called Hudson River formation, which crops out to the east of the Blue Stone area and forms in part the Taconic Mountain range along the New York-Massachusetts boundary line. This rock is of much greater age than the "Blue Stone," belonging to the Ordovician period of the geological series.

Similar examination of this Ordovician elastic rock shows that it in turn was derived from a still older elastic rock, this time of Pre-Cambrian age, some of which can be seen in the ledges of Manhattan Island. This elastic rock finally was derived from still older igneous and metamorphic rocks.

A second example from the central region of our country may be given. In Ohio, Michigan, and Western Ontario is a great bed of very pure sandstone, a elastic rock made up almost entirely of small, well-rounded grains of quartz sand bound together. This rock is so pure that it is used for glass-making in Toledo, Ohio. It is known by the name of the *Sylvania sandstone,* and it belongs to the Silurian division of the geological series.

Careful study has shown that the grains were originally a part of a still older sandstone, called the Saint Peter sandstone, which belongs to the Ordovician division and covers a large area in the Mississippi Valley region and cast of that in Michigan, Canada, etc., though its outcrops are found only in restricted areas. This rock was in turn derived, at least in part, from the destruction of a still older sandstone, the so-called Potsdam sandstone of Cambrian age, which crops out farther to the north, in Wisconsin, Minnesota, etc.

This sandstone finally was derived from the still older granites, gneisses, etc., which are seen in the Canadian region to the north. Thus the Sylvania represents the third generation of sandstone, the Saint Peter the second, and the Potsdam the first.

Finally we may note that the modern sands of the Libyan desert are derived from the breaking up into its component grains of an older sandstone, the Nubian, and of the rock from which the sphinx has been cut. If these sands become bound together into a sandstone at some future time, this sandstone will be of the second generation.

That elastic rocks may be of different generations. The first generation is always derived from some igneous or other non-elastic rock or the metamorphosed product of such a rock. The later generations are derived successively from older elastics, though new contributions from the crystalline source may also be made. This formation of a new and younger elastic rock from the material of an older one may be compared with the construction, in more recent times, of man-made structures and monuments from the stones obtained by the demolition of older human monuments or structures.

The cycle of change which includes the successive generations of a elastic series may be brought to an end by the melting of the elastics, and by their incorporation into a new igneous mass, or by their pronounced metamorphism. Then a new cycle of formation of elastic rock will begin.

THE AGENTS ACTIVE IN THE BREAKING UP

We may now consider the chief agents which are active in the breaking up or clastation of rocks, that is, those responsible for the formation of elastic material from which elastic rocks are formed by consolidation. We may classify the several types of elastic rocks according to the agent which produces the material, or which arranges it in the form in which it becomes consolidated.

The Atmosphere as a Rock Breaker

In the first place, rocks are broken up or "clastated" by the atmosphere acting either chemically, by the action of the various gases and vapors of the air upon the rock, or mechanically, as in the case of freezing moisture, or by the wind. Clastic material thus produced may be called *atmoclastic* material, and rocks formed by the consolidation of such elastic material may be called *atmoclastic rocks*.

As elastic material is often accumulated in certain localities after transportation by the wind, *i.e.*, the atmosphere in motion, and as this material has generally a very definite form and structure, we may further distinguish wind-arranged or wind-deposited elastic material, such as is found in sand-dunes. When this is consolidated into a rock it becomes a wind-formed or *eolian* rock, a rock which may also be called *anemoclastic* from the Greek *anemos,* the wind.

There are many examples of such wind-formed, eolian, or anemoclastic rocks to which we shall call attention. We may mention here as examples the recent dune-rock of the Bermuda Islands, and the much older sandstones of the White Cliffs in the Colorado Plateau region. The first of these was formed in the modern or *Holocene* period, the second in the Jurassic period, and this has retained its structure ever since that time, though much of the original rock, which had a far wider distribution, has been worn away again from large areas.

The Hydrosphere as a Rock Breaker

The second great agent which accomplishes the destruction of rocks is the hydrosphere. This destruction, so far as it bears on our present point of view, is mechanical in its nature, although chemical work, the solution of rocks (such as limestone, salt, etc.) is also performed by the hydrosphere. Such solution, however, results in the reincorporation of the rock material into the material of the *hydrosphere* from which in turn it may be deposited as an aqueous or hydrogenic rock (stalactites and other cave deposits). Such a rock is, however, not a elastic but a "genie" rock, and the student should learn to make the proper distinction between these two.

The mechanical work of water is manifested in streams, where the current moves the rock fragments which it has broken from the ledges, and grinds them down as it carries them along. It is also seen on the seashore, where the waves produce an analogous effect. But in addition to the breaking off of the rock fragments and the grinding of them to sand and pebbles, moving water in streams and on the sea-coast assorts elastic material, however produced, into grades of various sizes, and more or less according to the weight, and therefore the nature, of the material.

Moreover, elastic material deposited by and in the water will generally be characterized by a bedded structure; that is, it will exhibit, if seen in section, a succession of layers or strata one above the other, each one of which was at one time the topmost layer. Such deposits are called *stratified,* and of course, at any one time, only the top of the topmost *stratum* is visible. Where a river, however, has cut a gorge into an older stratified series, or where the waves on the sea-coast or lake shore again partly wear away such a series which has been lifted above the level of the sea by natural disturbances, or rises above the lake-level, the cut edges of such strata can be seen. Thus on the opposite banks of the Genesee River we see the cut edges of the successive strata which were formerly continuous across the chasm.

The Pyrosphere as a Rock Breaker

The pyrosphere contributes pyrogenic or igneous material, which on cooling and hardening forms a part of the solid crust of the earth.

It also breaks up material already rock, and this is generally brought about by explosive activities such as are characteristic of most volcanoes.

By these activities the rock, whether an older lava or a rock of other origin, located in and around the crater, may be shattered and the material thrown high into the air, to descend as a shower of ashes or larger particles. Such material, which is readily recognized by its character, is called *pyroclastic* material, and when bound together to form a rock, this becomes a pyroclastic rock. Examples: volcanic tuff, volcanic agglomerate, etc.

Shattering of Rocks within the Lithosphere by Movements

When one part of the lithosphere moves over or against another part, as happens when the earth's crust suffers disturbances the rock along the plane of movement is shattered or ground fine, and clastic material with very definite characteristics is produced. As such a movement of adjoining rock masses generally produces a displacement, or fault, the shattered or ground-up material is commonly called a fault-breccia. A more general term for such material is autoclastic, because it is produced by the self-destruction or breaking up of the rocks of the earth's crust.

The Biosphere as a Rock Breaker

That growing plants, such as trees, arising from. seeds which lodged in a fissure of the rock, can by expansive growth shatter that rock, has been frequently observed. Animals, too, break rocks. Thus where a spring issues upon a level surface in a more or less arid country, vast herds of hoofed animals will congregate to drink, and their constant stamping of the rock surface will break it and produce dust and sand which mingles with the water and which on drying may be carried away by the wind.

Thus hollows are produced in the surface of the land, and these may be filled with water and so produce ponds. Many such are known to exist in western North America, in north and central Africa, and elsewhere. Fish and other animals in the sea will break off branches of coral from reefs (biogenic rock) and grind these to powder to obtain for nourishment the animal matter which surrounds these coral masses. Thus much coral sand and mud is produced, and this is clastic rock material of strictly organic origin.

But by far the greatest destroyer of rock is man. Rock is broken down to be used for the making of artificial rock, that is man-made rock (rubble-concrete, etc.). As man is a part of the organic world or biosphere, his work must be classed with that of other organisms. Clastic material thus produced by organisms, and the rocks, whether natural or artificial (i.e., man-made), made from these, may therefore be called *bioclastic*.

Chapter 4

Pyrogenic Rocks

THE IGNEOUS MAGMA

The term molten magma is applied to molten rock material or igneous fluid, which is formed or exists within that part of the earth which we have called the pyrosphere, and which, it will be remembered, interpenetrates the lithosphere and the asthenosphere (tectosphere). When the magma reaches the surface of the earth, either in a volcanic eruption, or through large fissures in the earth's crust, as in parts of Iceland to-day, it is called a *fluid lava*, from which by solidification in cooling a *solid lava* or lava-rock is produced.

Such a rock mass constitutes an *extrusive* or *effusive* igneous rock mass. When the magma does not reach the surface, but cools within the crust of the earth, into which it has risen or become intruded to a greater or less extent, either upon or by the formation of fissures, or otherwise, it forms an *intrusive* igneous mass. Finally, a magma may be conceived as cooling essentially where it was formed, and to produce a *deep-seated* or *abyssal* igneous rock mass.

Formation of Igneous Rock

It is obvious that only the surface-flows or lavas will be accessible to man immediately after cooling, and of these generally only the superficial portion. All the other igneous rock masses are located within the earth's crust, and buried beneath the surface rock masses, sometimes to very great depths. Some of those which closely approach the surface may be reached by deep borings or by mining operations, but this can occur only in exceptional cases.

Some of these may also be conceived of as brought into view on the face of a great dislocation-block of the earth's crust, where one side of the broken crust is lifted and the other depressed. This too, however, is probably so exceptional that it may be considered, in the absence of other modifications, to be practically negligible. In general, it is only after prolonged erosion, which results in the removal of much of the surface-covering of the rock, that intrusive masses become exposed, while deep-seated masses may require the removal of many thousands of feet of covering rock before they become visible.

This removal of great covering rock masses requires, of course, long periods of time, and it thus becomes evident that the igneous rock masses, other than surface lavas, which are now visible in outcrops, are of great age, and that their formation by cooling has taken place at remote periods of time.

Igneous intrusions and deep-seated masses which are now being formed are entirely invisible to us, and will not be exposed until many ages have passed by, if ever. Nevertheless there are indirect ways in which we can infer the existence of igneous masses beneath the surface, which are now undergoing the process of solidification, and to some of these we shall refer later.

Characters and Composition of Igneous Magmas

It is not possible to gain a complete knowledge of the composition of an igneous magma from the composition of the igneous rock which has resulted by solidification from that magma, because the magma contains in addition to the substances which make up the solid rock formed from it, large quantities of volatile gases which are expelled upon cooling and relief of pressure.

The volatile gases which are expelled under these conditions are, water vapour, carbon dioxide, hydrochloric acid, sulphurous vapors, and the like. Such expulsion of volatile matter is shown by all surface lavas from which clouds of steam arise, and such steam on analysis proves to carry with it many of the other gaseous emanations. Others quickly condense upon the surface and may form salts of various kinds, which either encrust the surface of the cooling lava, or are carried away in solution by the condensing waters from the water vapors. Sometimes the expulsion of gas and vapors is so rapid that violent explosions result, and this is indeed the chief cause in the production of pyroclastic material. In practically every volcanic eruption there is produced a cloud of vapour and gases which carries upward vast masses of finely divided rock material — the product of the explosive action — and often rises to heights of many miles above the volcano. When gases and vapors alone issue from a fissure in the earth, the phenomenon is spoken of as a *fumarole*. Fumaroles, are commonly associated with declining volcanic activities. The gases and waters of cooling magmas within the earth's crust may escape through fissures not penetrated by the magma or opened since cooling began, and these may reach the surface as mineral springs, either hot or cold. In their upward passage, such waters and gases may deposit mineral matter which they carried, and this is believed by many to account for vein and other mineral deposits.

The non-volatile material of the magma which solidifies to form the lava or other igneous rocks consists predominantly of only a small number of substances, chief among which is silica (SiO_2), which, however, is present in variable amounts, according to the nature of the magma. In addition to this there are the oxides of the metals aluminum (Al), iron (Fe), magnesium (Mg), calcium (Ca), sodium (Na), and potassium (K). These, too, vary in amounts in the different magmas. On solidification the silica unites with them to form various *silicate* minerals, which in the coarser-grained igneous rocks can be readily distinguished. In a general way there can be recognized a gradation in the composition of the magmas from a point where silica is most abundant, sometimes forming 75 per cent of the whole mass, with alumina and potassium next, by an increase in sodium and later in calcium, magnesium, and iron, and a reduction in silica to a point where this constitutes 50 per cent or less (rarely 135 per cent) of the entire mass. With this occurs a reduction in the oxides of potassium, sodium, and aluminum, sometimes to the complete or nearly complete elimination of some of these. The end high in silica, etc., is called the *acid* end of the series; the other end is the *basic* end. Rocks formed from the acid portion of an igneous magma generally have light-colored and light-weight minerals predominating; those formed from the basic portion have mainly dark-colored and heavy minerals.

Acid magmas (and lavas) are generally very stiff or *viscous* even at high temperatures (2000° C. or over), and their contained gases escape with difficulty, and often with explosive violence, as the magma approaches the surface. Basic magmas (and lavas), on the other hand, are more fluid even at much lower temperatures (1300° C.), and on this account the gases escape more readily, and explosions are less common. The lavas of Vesuvius and Mont Pelée are examples of the first, those of Kilauea in Hawaii, of the second.

FORMATION OF IGNEOUS ROCKS BY COOLING OF MAGMA

With the cooling of the non-volatile part of the magma, solidification takes place and igneous rocks are produced. The kind of rock will of course vary with the variation in the composition of the magma, which must be considered to be the character of first importance. Next to this is the rate of cooling, which determines the grain or texture of the rock, and which in turn is influenced by the relative position of the magma while cooling.

INFLUENCE OF RATE OF COOLING

It is a well-ascertained fact that a phyritic texture.rapidly cooling magma will tend to produce a mass of glass, and that in proportion as the rate is slower will there be opportunity for the growth of mineral crystals, the size of which is, in general, proportional to the slowness of the cooling. When the entire rock becomes a mass of mineral crystals, it is said to be *holocrystalline*. Between this and the glassy type in which there are no crystals, there are all gradations. The relative size and arrangement of the component crystals of a rock form its *texture*.

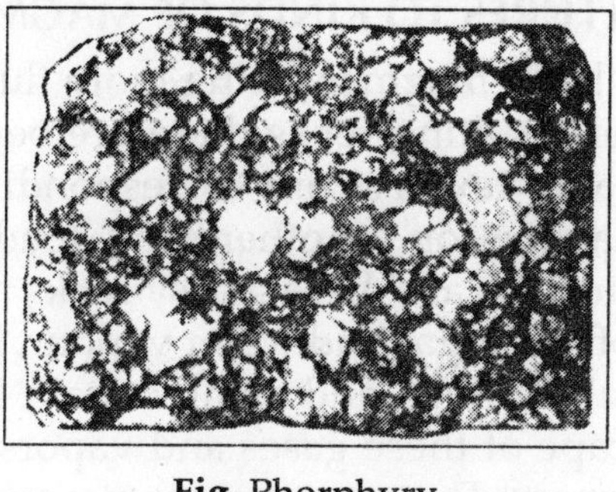

Fig. Phorphyry

Primary Textures

Two general types of primary texture may be distinguished, the *homogeneous* or *uniform*, and the *heterogeneous* or *porphyritic*. In the first all the crystals of each mineral are essentially of uniform size, though the size varies with the mineral. In the second, or *porphyritic* type, on the other hand, the crystals of certain of the essential minerals (generally the feldspars) are of two sizes, one large and more or less fully and perfectly formed, the other small and less perfect.

The larger crystals, which are scattered through the mass, are called the *phenocrysts*; the finer-grained mass, together with other minerals, also in small grains, forms the *ground-mass*.

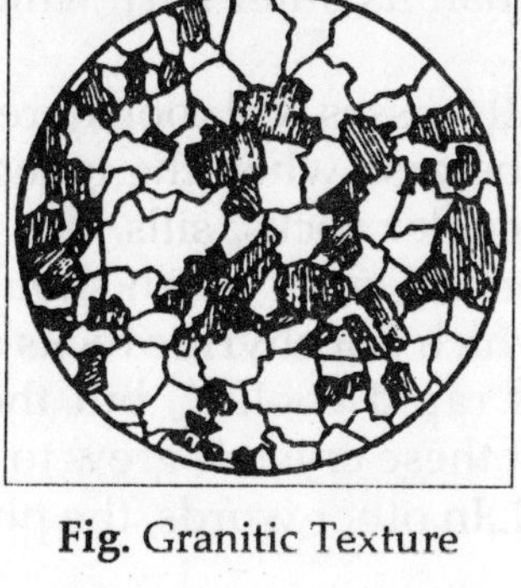

Fig. Granitic Texture

Secondary Textures

The textures of the ground-mass and of the homogeneous textured rocks are called secondary. When the grains or crystals are all nearly uniform, the texture is said to be *granitic* or *granular,* and it may be coarsely or finely *granular, i.e.,* the texture is *coarse-grained* or *fine-grained* so long as the individual crystals are distinguishable. But when the crystals are no longer distinguishable, the texture becomes *dense* or *felsitic,* while a step further brings us to the *glassy* texture in which no crystals are developed.

RELATION OF TEXTURES TO KINDS OF MAGMAS

In general, the basic magmas, being more fluid, will tend to form coarser crystals, the resulting rocks therefore being more commonly coarse-grained. Glassy textures are correspondingly less frequently developed. Acid magmas, on the other hand, tend to develop the finer textures more frequently, and here glassy rocks are common.

The presence of much gas and water vapour, too, tends to increase the power of crystal forming, and along certain fissures, presumably the pathway of escape of these gases and vapors, the texture is often of exceeding coarseness. The most common examples are the dikes or other masses of *pegmatite,* a rock composed sometimes of huge crystals of feldspar, quartz, and mica, the latter mineral being obtained in large plates from this rock, which is the chief commercial source of this mineral.

BULK OF MAGMA

As the coarseness of grain is in proportion to the slowness of cooling, it is evident that, other things being equal, whatever lowers the rate of cooling will cause an increase in the size of the crystals, and of the coarseness of texture. Thus a magma cooling within the earth's crust, especially at great depth, will of necessity cool slowly, and so become a coarsely crystalline rock. Again, the central part of the magma will cool more slowly than its outer part, which is in contact with the cooler enclosing rock.

In like manner small masses will cool more rapidly than large ones, and this is especially the case when the igneous mass is a thin sheet transecting other and cooler rocks, sills, etc. There is, however, one modification of this general rule, which should be noted, and that is the conditions under which porphyritic rocks appear. In these the fine ground-mass indicates rapid cooling, but the presence of the large phenocrysts shows that these crystals grew to their full size before the ground-mass solidified. In other words, the phenocrysts were floating crystals in a still semi-fluid matrix. Typical porphyries (with fine-grained ground-mass) are most common in certain lava flows, somewhat less so in the smaller intrusive masses, and still less common, and indeed rather rare, in the great subterranean masses. All types of igneous rocks may, however, show a porphyritic texture.

CLASSIFICATION OF IGNEOUS ROCKS

From what has been said up to this point, it appears that the principles which underlie the classification of igneous rock are relatively simple.

The first general division is made upon a chemical basis, which in the rocks is expressed by the mineral species present. Further subdivisions are based upon the texture of the rock, for the same magma may furnish rocks of different textures as the result of cooling at different rates under different conditions.

Since, as above stated, the character of the magma is most readily ascertained from the minerals which crystallize from it, chief attention is ordinarily given to these minerals. Furthermore, since these are readily recognizable only in the coarse-grained or holocrystalline products of cooling of the magmas, it is customary to designate the rock groups based on chemical and mineralogical characters by the name of the typical holocrystalline member of each group. Before we consider these types, however, we must briefly review the essential minerals which enter into the construction of these rocks.

The Essential Minerals of Igneous Rocks

We may in general distinguish three groups of essential minerals which make up the bulk of the igneous rocks. These are the following, arranged in each case in the order from acidic to basic. They are:

- Quartz,
- The Feldspars and Feldspathoids
- The Micas and Ferro-Magnesian Silicates.

Quartz (SiO_2)

This generally occurs in glassy fragments of irregular and usually of angular form, more rarely in crystals. It is easily recognized by its hardness.

The Feldspars

In composition these are all double silicates of aluminum and an alkali metal (K, Na, etc.) or an alkaline earth (Ca, Mg, etc.) or both. In general the feldspars, are divided into *orthoclase* or the feldspar with right-angled cleavage, and *plagioclase* or the feldspar with oblique cleavages.

The most acid feldspar contains only potassium in addition to both alumina and silicic acid (potash feldspar or orthoclase), the most basic only lime in place of the potash (lime feldspar or anorthite). Between the two stands the soda feldspar (albite) where sodium is the metallic base besides aluminum. Between the pure soda and the pure lime feldspars are a number of intermediate types, consisting of different proportions of both soda and lime; and between the pure potash and soda feldspars there are also mixtures of the two.

These relations may be expressed in the following table, where the three main or (theoretically) pure types of feldspars are each expressed by a molecular symbol and the mixed types by formulæ which indicate the proportions of each.

The Feldspathoids

Under this designation are placed double silicates of aluminum and the alkalies or alkaline earths, which have many characters in common with the feldspars besides the general similarity in their composition, but differ in crystal form and some other characters.

Theoretically we again have three types, the potash, soda, and lime feldspathoids, but actually the first two are more commonly intermixtures. This is shown in the following table. In the order of their importance as rock constituents, these minerals are *nephelite, leucite,* and *melilite,* the last being very rare.

Table. Feldspathoids

MINERAL **(pure)**	**COMPOSITION** **USUAL MODIFICATION**
Leucite or $KALSi_2O_6$ part of K_2O	$K_2O \cdot Al_2O_3 \cdot 4SiO_2$ Some Na_2O replaces
Nephelite or (almost) $NaAlSiO_4$ replaces part of Na_2O	$4\ Na_2O \cdot 4Al_2O_3 \cdot 9SiO_2$ Some K_2O and CaO
Melilite or $Ca_3Al(SiO_4)_3$	$12CaO \cdot 2Al_2O_3 \cdot 9SiO_2$

In general, the occurrence of leucite in igneous rock implies a magma rich in potash; nephelite, one rich in alumina and soda; and melilite, one poor in silica and alumina but rich in lime. Melilite occurs mainly in a few rare basalts.

Common Mica and the Ferromagnesian Minerals

The name ferromagnesian silicates is given to the minerals (usually dark in colour) which include in their composition both iron and magnesium. They comprise the dark mica, *biotite*, the *hornblendes*, *pyroxenes*, and *olivine*. The common or white mica (*muscovite*) has neither iron nor magnesium, but is a silicate of potassium and aluminum together with some hydrogen. Its usual formula is $K(OH) \cdot Al_2O_3 \cdot 2SiO_2$, or $HKAl_2\,Si_2O_8$ and it is commonly called the potash mica. As might be inferred, it occurs in rocks where other potash minerals are common, such as granites, pegmatites, etc. It is also common in metamorphic schists.

In the ferromagnesian silicates we trace a gradation in composition from biotite, or black mica, in which potash is present, through the amphiboles and pyroxenes, in which the potash is replaced by lime, to olivine, where no alkali or alkaline earth is present. This is shown in the following table where muscovite is also placed at the acidic end and magnetite, the pure iron oxide, at the basic end.

Table of the Ferromagnesian Minerals and Their Two End Members

Minerals	Composition	Usual Modification
Muscovite	$H_2\,KAl_3\,(SiO_4)_3$	
to		
$(HK)Al_2\,(SiO_4)_2$	Without iron or magnesium.	
Biotite	$(HK)_2\,(MgFe)_2\,Al_2\,(SiO_4)_3$	
(nearly)	Addition of magnesium and iron;	
	reduction of alumina and silica.	
Amphibole		
(Hornblende)	$Ca(MgFe)_3\,(SiO_3)_4$	
(in general)	Substitution of calcium for hydrogen,	
present.	potassium,and aluminum, though	
	some of the last may be	
Pyroxene		
(Augite)	$Ca(MgFe)(SiO_3)_4$	
(in general)		
Olivine	$(MgFe)_2\,SiO_4$	Omission of calcium,
		some reduction of silica.
Magnetite	$Fe_3\,O_4$	Omission of silica and
	magnesium.	

It must be understood that the amphiboles and pyroxenes are more complex in composition than here stated, and that there are a number of varieties of each, differing in composition. It is, however, not necessary that these be considered here. The common amphiboles and pyroxenes crystallize in the monoclinic system, but there are also orthorhombic members of each group. The chief means of distinction is the (prismatic) cleavage form, that of the pyroxenes being nearly at right angles, and that of the amphiboles forming angles of nearly 120 and 60 degrees.

In general, it may be said that whereas muscovite occurs chiefly in the most acidic rocks, biotite and hornblende indicate greater basicity, pyroxenes still greater basicity, while olivine is characteristic only of the very basic igneous rocks. When pyroxene and olivine are present, free quartz is usually absent.

Secondary or Accessory Minerals

There are many minerals which occur in small quantities in igneous rocks but are not necessary constituents of them. They are called accessory minerals, and their presence is most frequently detected by the microscope. Zircon and titanite are good examples of these. By alteration many other accessory minerals are formed from the primary or original ones.

Order of Crystallization

In general the order of crystallization of the minerals from an igneous magma follows the order of their basicity, the most basic separating out first, the most acid last. Thus olivine will form in perfect crystals from a basaltic magma, the other minerals being less perfectly crystallized. The pyroxenes and the hornblendes separate out before the feldspathoids and the feldspars, while the quartz, if there is any free silica remaining, separates out last, filling the interstices between the other minerals.

For that reason quartz is practically never in perfect crystal form in granites or other igneous rocks which contain it, but forms irregular grains, the shape of which is determined by the form of the cavities left between the other minerals. It has a crystalline structure, however, though not a crystal form.

TYPES OF IGNEOUS ROCKS BASED ON COMPOSITION

The more important types of igneous rocks, beginning with the most acidic on the left, and extending to the basic types on the right. It will be seen that there are certain composition groups based upon the kind of feldspar (or feldspathoid) present, the kind of ferromagnesian mineral, and the presence or absence of free quartz or olivine. Each of these composition groups includes a series of rocks ranging in texture from coarse (at the bottom) through fine and dense or felsitic to glassy (at the top).

The coarser-grained rocks are generally formed as deepseated masses, which have become exposed by erosion. The intermediate types occur mainly as intrusive masses, while the dense and glassy types (including the cellular) are primarily formed from surface flows. As the mineral character of the several groups is most readily ascertained in the case of the coarse-grained varieties, the various groups will be considered under these respective types. Microscopic examination is generally necessary to ascertain the characters of the finer-grained and dense types, though typical examples may generally be recognized without such aid.

THE GRANITE-RHYOLITE SERIES

This contains the most acidic of the common igneous rocks. Common types are: *granite, rhyolite, quartz-felsite, obsidian,* and *pumice,* and the porphyritic varieties of these.

Granite

The coarse-grained members consist primarily of orthoclase feldspar and free quartz. When these alone are present, as is sometimes the case, the rock is called a binary granite. Generally, however, black mica (biotite) and often muscovite and some hornblende are present, these being readily recognized by their dark colour, while the biotite is distinguished from the hornblende by its scaliness and softness. The orthoclase is commonly recognized by its pinkish or flesh colour and by the smooth cleavage planes which it shows, as well as the more or less marked crystal form. The quartz is always in irregular, glassy masses, breaks with rough fracture and, as a result, looks darker than the feldspar by reflected light.

Granites vary by the replacement of some of the orthoclase by acid plagioclase while hornblende may increase in amount, forming a *hornblendic granite.* With this may go a reduction in the quantity of free quartz present, when the rock approaches a syenite in composition, while with the increase in plagioclase the quartz-diorites or diorites are approached.

In rare cases, too, augite may occur, showing an approach to gabbro. Among the common accessory minerals are magnetite, zircon, and garnet. A porphyritic texture is sometimes developed by the formation of large and more or less perfect crystals of orthoclase, which frequently show a twinned character (Carlsbad twins), recognizable by the fact that one half of the crystal appears darker than the other in reflected light, the shade being reversed with change in the position of the light.

Pegmatite

This is the name given to a coarse variety of a granite which occurs in vein or dike-like masses, and generally consists of large crystals of orthoclase (sometimes acid plagioclase) occasionally up to a foot or more in diameter, large masses of quartz, and large plates of mica (muscovite).

The quartz and feldspar are sometimes found intergrown in such a manner that the surface of the feldspar seems to be scattered over with small, irregular, dark masses of quartz, appearing not unlike cuneiform characters. On this account such a mixture of feldspar and quartz is called *graphic granite.*

Rhyolite and Quartz Felsites

These have essentially the same mineralogical composition as granite, though the association may vary more, producing greater variety. In colour the rocks are gen erally light gray, with yellows, and pale reds, and occasionally darker shades. The texture varies considerably, from very finely crystalline to dense or felsitic, when the rock consists of minute masses of quartz and feldspar often with more or less glass. A cellular structure may also be developed to a slight degree in rhyolites formed from surface flows. In many rhyolites, especially if they are porphyritic, quartz can be recognized (commonly as small double-sided pyramids), giving the surface a rough feel.

Phenocrysts of feldspar also occur in the porphyritic types. When the phenocrysts of quartz and feldspar make up about half the mass of the rock it is called a *rhyolite-porphyry;* when it constitutes most of the rock it is called a *granite-porphyry* — and marks an approach to granite. In both the ground mass is generally dense or felsitic. The name rhyolite is given to this rock because of the flow structure often exhibited.

The Acid Glasses

These include obsidian, pitchstone, perlite, and pumice. In these the excess of silica, which on crystallization would produce free quartz, can generally be recognized only on analysis, except when phenocrysts are developed. As the magmas which produce syenites on crystallization also form glasses of similar appearance, though poorer in silica, it is evident that it is not possible to determine, except by analysis, whether a given glass belongs to that or to the granite series.

Obsidian

This is a homogeneous glass with a low percentage of water. It is black to red in colour, with translucent edges, and a conchoidal fracture, so called because the surface of the fracture is generally marked by a series of concentric lines, like the growth-lines of a shell.

Pitchstone

This is like obsidian, but contains from 5 to 10 per cent of water. It has commonly a sheen or luster suggestive of resin, and its colors are commonly reds and greens.

Perlite, or Pearl Stone

This is glass composed of numerous rounded masses, with concentric structure like the coats of an onion, formed by contraction in cooling and separated by larger straight cracks. It usually contains from 2 to 4 per cent of water.

The Syenite-Trachyte Series

This series is primarily characterized by deficiency in silica, so that no free quartz is formed on crystallization. It includes syenite, trachyte, felsites, and glasses.

Syenite

This rock typically consists of orthoclase and hornblende, without quartz. When biotite is present, the rock is called *mica syenite*. In practically all syenites some of the orthoclase is replaced by plagioclase, and this may occur to such a degree that it makes up half of the feldspar, when the rock approaches a diorite, and is called a *monzonite*. Augite too may replace the hornblende, occurring with orthoclase and forming an *augite syenite*.

Finally, *quartz syenite*, with a small amount of free quartz, shows a transition from granites. With the appearance of nephelite, they pass into *nephelite syenites*. Porphyritic syenites have large crystals of feldspar. Common accessory minerals are magnetite, zircon, and apatite. The name syenite is derived from the ancient Syene in Egypt, where the rock is, however, a hornblende granite which was formerly used for obelisks. As now used the name was first applied to a granite rock almost without quartz near Dresden.

TRACHYTES AND FELSITES

These have essentially the same composition as the syenites, of which they form the crypto-crystalline representatives with the same relationship that rhyolite holds to granite. Trachytes differ from rhyolites in the absence or great rarity of quartz. Biotite is in general the most abundant dark mineral, but hornblende and augite also occur, forming varieties.

The texture varies from felsitic in the true felsites to strongly and coarsely porphyritic, and not infrequently the rock is somewhat cellular. Felsites, on account of their dense structure, cannot readily be distinguished from rocks of the same type in the granite or even in the diorite group. In typical trachytes, which are generally porphyritic, the ground-mass is made up of fine rods of orthoclase arranged more or less parallel and in flowing lines.

This is the characteristic trachyte texture, which can often be seen with the naked eye in the coarser crystalline varieties. Large phenocrysts of a clear vitreous variety of orthoclase are characteristic of the porphyritic trachytes.

When the phenocrysts are so abundant as to constitute about half the mass of the rock with felsitic ground-mass, it is called a *trachyte porphyry*. When phenocrysts form the bulk of the rock, while the ground-mass becomes more coarse-textured, the rock is called a *syenite porphyry* and marks the transition to syenite.

The name trachyte is derived from its rough or harsh surface feel. The most typical trachytes come from the peak of the Drachenfels on the Rhine.

The Glasses of the Syenite Series

Because the fusing point of these more basic rocks is lower than that of the granite series, being about 2000° F. (1100° C.) for trachytes as compared with about 2200° F. (1200° C.) for the rhyolites, and 2250° F. (1240° C.) for granites, they remain liquid longer, and hence crystallization occurs more generally with less frequent formation of glasses.

THE NEPHELITE-SYENITE PHONOLITE SERIES

These differ from syenites chemically in the greater amount of soda, and mineralogically in the partial substitution of nephelite (eleolite) for the feldspar.

Nephelite-Syenite

This corresponds to syenite except for the presence of nephelite or sodalite, of both or of leucite. In some cases (Litchfield, Maine) the feldspar is wholly plagioclase. Zircon is a usual secondary mineral.

Phonolites

These rocks are generally dense and finely crystalline, seldom vesicular or glassy. They are mostly dull green or gray in colour, and when light colored they are not readily distinguished from trachytes except by the microscope, which shows the presence of the nephelite. The chief feldspar is orthoclase, while the common dark mineral is augite, hornblende being rare. A peculiar character is the fact that the rock breaks into thin slabs which have a musical ring under the hammer.

The rock is frequently porphyritic, both augite and feldspar phenocrysts appearing in a dense ground-mass. When the phenocrysts are very abundant the rock is called *phonolite-porphyry,* and when in excess, it becomes a *nephelite-syenite porphyry.* With increase in orthoclase and decrease in nephelite, the phonolites pass into trachytes.

Glasses of the Nephelite-Syenite Series

These are still rarer than those of the preceding case, since the fusing point of phonolites (somewhat less than 2000° F. or 1090° C.) is still lower than that of trachytes. Phonolite obsidians are known from the Peak of Teneriffe.

THE QUARTZ-DIORITE DACITE SERIES

The members of this series differ from those of the granite series, chiefly in the substitution of plagioclase for orthoclase. The plagioclase can be recognized by the striated character of the cleavage surface.

Quartz-Diorite

This resembles granite, but is darker and heavier. Acid plagioclase, quartz, hornblende, and (or) biotite are the essential minerals. When biotite predominates the rock is called a *quartz-mica-diorite.* These rocks form a transition from granites to diorites, the intermediate forms being called *granodiorites.* A typical locality for quartz-mica-diorite is found in the Cortland Series near Peekskill, N. Y., one for the hornblendic quartz-diorite in the Yellowstone Park.

Dacite

This rock is difficult to distinguish from rhyolite except by the use of the microscope. It has the same relation to quartzdiorite that rhyolite has to granite. When the texture is finely felsitic and non-porphyritic, it can only be termed felsite. When porphyritic, the dacites are recognized by the striated surfaces of the plagioclase phenocrysts, which predominate.

Glasses and cellular texture are not uncommon. When phenocrysts constitute up to half the mass of the rock, it becomes a *dacite-porphyry*. When they are in marked excess over the ground-mass, it becomes a *quartz-diorite-porphyry*. The name is derived from the old province of Dacia, now Transylvania (Siebenbürgen), before the war a part of Hungary, which is a typical locality. Dacites generally occur with andesites.

Glasses of this Series

The glasses of this series are more common than those of the syenite series, for the fusing point is only a little less than that of the granite series. They are generally included with the glasses of the next series.

THE DIORITE-ANDESITE SERIES

This differs from the preceding in the absence of free quartz (in the crystalline members) and from the syenite series in the substitution of acid plagioclase for orthoclase.

Diorites

These are granitoid rocks, the chief feldspar of which is acid plagioclase, and they are rich in hornblende. When biotite largely replaces the hornblende, the rock is called *mica-diorite*. Again, augite may replace part of the other dark minerals, forming an *augite diorite*, which is a passage-rock to gabbro. Typically, the feldspar is in excess of the dark minerals, but in other cases these may lead in the mineral constituents.

A rock of diorite composition may also arise by metamorphism of gabbros with the change of the augite to hornblende. Porphyritic diorites, occurring in dikes, have been called *camptonites*. Characteristic secondary minerals are: magnetite, titanite, and apatite. The name *diorite* was given to this rock because of the striking contrast between the light and dark minerals of which it is composed.

Andesites

These are the fine-textured to felsitic members of the diorite series. The acid plagioclase feldspars are the most abundant minerals, but quartz is rare or absent. Biotite, hornblende, and augite are the dark minerals, the last two predominating over the biotite. The general colors of the rocks are grays or greens. Typical andesite is felsitic, sometimes cellular, and commonly porphyritic.

The felsitic ground-mass consists of microscopic rods of feldspar, forming a felt-like aggregate. From trachyte and dacite it can generally be distinguished only by its darker colour, owing to the greater abundance of the ferromagnesian silicates. According to the predominant dark mineral we have mica andesite, hornblende andesite, augite andesite, etc.

In the porphyritic varieties, the phenocrysts are mostly feldspar; when very abundant (up to half the mass) they form an *andesiteporphyry;* when in excess, a *diorite-porphyry*. These show an increasingly coarser ground-mass, and grade into diorites proper. With increase in orthoclase, andesites pass into trachytes, and with increase in the dark and more basic ferromagnesian silicates and a decrease of feldspars, they pass into basalts; with addition of quartz they pass into dacites.

The type localities for these rocks are in the Andes Mountains, and they are abundant and widespread in the Pacific Coast region of North America, especially in the old volcanic cones of Mt. Hood, Mt. Shasta, Mt. Rainier, and others.

The Glasses of the Diorite-Andesite Series

Andesites fuse at temperatures around 2000° F. (1100° C.), which is about the same as for trachytes. Like these, therefore, they do not readily form glasses, and most of those referred to andesites are probably referable to the dacites. The glasses of this type which ate recognized are: andesite-obsidian, andesite-perlite, and andesite-pumice.

They are distinguished from the more acidic glasses only by chemical analysis, or by their field relations to recognizable types of andesites or dacites. Andesite obsidian has been obtained from Clear Lake, Cal., andesite perlite from Eureka, Nev.

THE GABBRO-(PYROXENITE, PERIDOTITE)-BASALT SERIES

This includes all the basic igneous rocks, the coarser-grained members of which range through gabbro, olivine-gabbro, pyroxenite and peridotites, while the fine-grained ones form diabases and basalt. The surface flows are represented by scoriaceous, ropy, or other lavas, and more rarely by glasses.

Gabbro and Olivine Gabbro

These are generally coarsely crystalline dark rocks, composed chiefly of basic plagioclase and monoclinic pyroxenes, the dark silicates typically predominating. There are, however, varieties which are almost wholly composed of coarse crystalline labradorite with little or no pyroxene. These are also called *anorthosites*; where the pyroxene is of the orthorhombic varieties (enstatite, bronzite, hypersthene), the rock is called *norite*.

Gabbro may contain some hornblende and biotite. When olivine is present the rock becomes olivine-gabbro, olivine-norite, etc. In rare cases nephelite becomes an important mineral in some basic gabbros, forming the rock *theralite*, which occurs in the Crazy Mountains, Montana. Gabbros have a wide distribution.

Pyroxenites and Peridotites

By the decrease of the plagioclase, the gabbros pass insensibly into pyroxenites and peridotites, which are usually found associated with the gabbros, but also occur independently. Pyroxenites generally contain little else than pyroxene, which may be orthorhombic (enstatite, bronzite, hypersthene) or monoclinic (diallage, augite), but these are not readily distinguishable by the unaided eye, though the orthorhombic pyroxenes show a bronze luster.

Frequent accessory minerals are hornblende, magnetite, and pyrrhotite. When olivine is added the rock becomes a *peridotite*, and when olivine predominates it becomes a *dunite*, of which the nearly pure olivine rock of North Carolina is an example. Sometimes magnetite is so abundant as to make the rock almost an iron ore. Porphyritic peridotites have been called *picrites*, especially when occurring in dikes. Sometimes hornblende is abundant, and it may even form a rock by itself, which is then called *amphibolite*. Pyroxenites and peridotites change with age and by metamorphism into serpentines.

DIABASE

This is a transitional rock between the gabbros and the basalts, being fine-grained but holocrystalline and differing in detail of texture from the gabbro. It consists of basic plagioclase, augite, and often olivine, the first occurring as elongated rectangular rods arranged in an interlacing manner, while the other minerals are packed in between them in irregular masses.

This is brought about by the fact that the plagioclase crystallized out first, contrary to the usual order of crystallization, and that the ferromagnesian minerals had to adapt themselves to the remaining spaces. This *diabasic texture* is sometimes modified into an *ophitic* one, where the rods of plagioclase are included in large, coarsely crystalline masses of pyroxene. Diabase is common as an intrusive rock, both as dikes and sills.

BASALT

These are the basic fine-grained or dense igneous rocks which form the surface flows of the gabbros, pyroxenites, and peridotites. Typically the texture is dense or felsitic, often cellular, and the rock may be characterized by almond-shaped steam holes which are subsequently filled with secondary calcite, giving the rock an *amygdaloidal* structure. Sometimes these cavities are filled with copper, as in the Lake Superior region. Por phyritic basalts show phenocrysts of augite and more commonly olivine. Such porphyritic olivine basalts are also called melaphyres. The felsitic ground mass is composed of microscopic augite, plagioclase and magnetite crystals. Sometimes a little glass is present. With abundant development of phenocrysts, of augite and olivine, the rock becomes a *basalt porphyry* and with excessive development of phenocrysts a *gabbro porphyry* is produced.

In rare basalts nephelite or leucite may replace the feldspar, giving a variety of types, distinguishable only under the microscope.

Augitites and Limburgites

These are rare basaltic rocks, with little or no feldspar, which correspond to the pyroxenites and peridotites, respectively.

The Basic Glasses

These represent surface formations of the basaltic flows, but are on the whole not common. Some of the porous or vesicular scoria is glassy, but for the most part it has a dense stony or felsitic texture. Typical basic glass, corresponding to obsidian, is known by the name of *tachylite,* while a fibrous variety found in the crater of Kilauea is called Pele's hair.

Special Field Names

Several names are commonly applied to the basic igneous rocks when the exact character cannot be determined in the field. One of these is *dolerite,* used for rocks which may be diorites or gabbros, but in which the dark mineral is undeterminable. Another is *trap,* in common use for dark, dense basalts or diabases. It is derived from the Swedish trappar, a stairway, because the sheets of this rock sometimes form successive steps in the landscape. Old trap or basaltic rocks which have, by alteration, developed sufficient of the mineral chlorite to give them a greenish cast, are called *greenstones.*

Chapter 5

Hydrogenic Rocks

Practically all natural waters contain mineral matter in solution, the common illustration being ocean water, every liter of which contains about 35 grams of mineral matter, more than 27 grams of this being common salt (sodium chloride), which means that every cubic mile of sea-water contains about 131,526,000 short tons of this important substance in solution. Sea-water is therefore spoken of as *salty* or *saline*, and this saltiness is readily recognized by taste.

The other dissolved substances are also called salts, but their presence is not so readily recognized. Some water bodies, like the Baltic Sea, contain only one fourth or less of this amount of salts in solution, and such waters are called *brackish*. The waters of the Hudson River some distance above New York City are brackish, because they are formed by a commingling of the fresh water from the upper river with salt water entering from the sea.

When the water contains so little substance in solution that it can be used for drinking purposes (whether it is contaminated by organic matter or not) it is called *fresh water*, but fresh water in nature always contains some substances in solution, lime usually predominating, this when present in sufficient quantity forming *hard water*. When carbonate of soda and similar substances are present in such quantities as to render the water unfit for human consumption, it is called *alkaline*, and all travelers in the semi-arid regions of western North America are familiar with such water.

Finally there are water bodies like Great Salt Lake, the Dead Sea, and others, in which the quantity of common salt in solution exceeds many times that in the ocean water, and such waters are spoken of as *super-saline*, or as *brines*. Brines may also be produced by partial evaporation of ocean water, just as brackish waters are produced by the dilution of ocean water.

SEPARATION OF MATERIAL

When the substances which waters hold in solution are separated out so as to form solid material, this material constitutes rock masses to which in general we apply the name of *salts*, after the example of the commonest of these, the ordinary salt.

If the separation is produced by organisms, the product is called organic salts, and these belong in the division of organic or biogenic rocks. If the separation is by inorganic activities, the product is a normal aqueous or hydrogenic rock is devoted.Separation by inorganic means is accomplished in a variety of ways, of which the following are the most important:

- Separation by the condensation or complete evaporation of the water by heat, drying winds, etc., during which process a stage is reached when the water becomes *saturated*, that is, it holds in solution as much of a given salt as it can hold for that temperature and pressure. If that stage is passed, the excess of salt separates out, often in a more or less crystalline form. The point of saturation varies with the nature of the salt, and when two or more salts are present in the solution, they will not only have separate and distinct saturation points, but the presence of each is likely to influence the saturation point of the others, either lowering or raising them. Complete separation of all the salts will occur upon complete evaporation of the water. Such salts are called evaporation products, or briefly, *evaporates*.
- Separation of salts from the solution by the force of attraction which crystals or particles of mineral matter exert on material of the same kind in the solution (generally a saturated one) in which those crystals or particles are immersed.
- Separation by the abstracting of the solvent or substance which holds the salt in solution.
- Separation by chemical reaction between minerals in solution in the water, and other substances introduced from extraneous sources, either as gases or as solutions, and the consequent formation of new and less soluble compounds which are then precipitated.
- Separation by electrolysis.

As illustrations of the first class we may cite the separation of salt when sea-water is evaporated or when incrustations of salt are formed from a solution of that substance which is allowed to stand for a time in a warm room. The second method is illustrated by the crystallization of the alum of a saturated solution around a crystal of alum introduced into that solution, or the formation of rock candy from a saturated solution of sugar. Precipitation through abstraction of a solvent is shown by the deposits of lime on the inside of boilers and tea-kettles where "hard" water is used, the solvent, which is carbon dioxide, being driven off by the heat. It is also shown by the precipitation of lime around the mouths of springs, in which the water escaping from under pressure loses some of the carbon dioxide which was the solvent of the lime in the water. Examples of precipitation of salts by the addition of substances in solution which by chemical reaction produce a less soluble compound, or by the passing of gases through a solution, are familiar to all workers in chemical laboratories, while the electrolytic method is one much practiced in the arts. The first three methods cited are most commonly observed in nature. The precipitation by the addition of reagents, whether liquid or gaseous, is chiefly found in the formation of lime deposits under the influence of ammoniacal gases.

Since these are, however, in nearly all cases produced by the direct activities or the indirect influence, through decay, of organisms, such deposits are best referred to the organic or biogenic group. Electrolytic processes in nature are still little understood, but that they are going on cannot be doubted. Some of the inclusions of salt in the pore spaces of marine sediments have been explained in this manner. Most precipitates from an aqueous solution are called salts, no matter how they are formed or what their composition. There are, however, some simpler substances, such as the oxide of silicon or silica (SiO_2) and others, which cannot properly be called salts. One of the essential characteristics of a salt is its purity of composition, though it is true that under certain conditions precipitation of several salts may take place simultaneously, thus producing what is called a *eutectic mixture*. When pure, the material is in reality a single mineral mass, as for example, the mineral Halite or rock salt; but because of its extensive development it is treated as a rock.

CLASSIFICATION OF AQUEOUS PRECIPITATES

It is evident that chemical composition is the primary basis on which aqueous precipitates must be classified. Furthermore, most salts look very much alike except when they are crystallized, or when, as in special cases, colour, hardness, and weight make distinction possible. In the following pages will be given some of the more important and common salts and oxides, the classification being made, for the sake of convenience, upon the basic element in the composition.

HYDROGENIC DEPOSITS

Among the many precipitates or other deposits formed directly from aqueous solutions, a certain number is found in sufficiently large quantities to be treated as rock material, while others are important as sources of valuable substances or are themselves of economic value.

NON-METALLIC AQUEOUS DEPOSITS

Rock Salt

Chloride of Sodium (NaCl)

This is the most important as well as the most abundant mineral substance obtained from ocean water. It commonly occurs in beds, which may have a thickness of a hundred feet or more but generally are only a few feet thick, forming a succession of beds separated by gypsum, anhydrite, limestone, dolomite, or clay or more rarely by other mineral matter.

Such deposits are formed by concentration of sea-water either in basins cut off from the sea by elevation or by the formation of a barrier, or in lagoons behind a bar with continued supply of sea-water through an inlet. Salt is also separated from lake basins in arid regions by the concentration, through partial evaporation, of the water. Finally, salt deposits are formed in desert basins from salt disseminated through the rocks in the rims of those basins.

Rock

Salt deposits are seldom very continuous, having in most cases a lenslike form. Salt is also extensively manufactured by the evaporation of seawater in shallow salt gardens or sea salinas, which are located on nearly all shores where the climatic and other conditions are favorable. Natural as well as artificial brines, formed by the solution of old salt beds in the earth's crust, or of disseminated salt, are also extensively used in the manufacture of salt. Salt is mined in central New York, in southern Michigan, in Louisiana, and in a few other localities in the United States.

Extensive salt mines exist in Austria, Galicia, Rumania, and elsewhere in the Old World. The salt mines of North Germany are worked chiefly for their potash deposits. Hills or mountains of salt are found in many of the dry regions of the world, — in northern Spain (Cardona), Algeria, Persia, and elsewhere. The chief uses are for domestic and dairy purposes and for chemical industries.

Gypsum

Hydrous sulphate of lime ($CaSO_4 \cdot 2\ H_2O$)

This is a common associate of salt deposits, being found beneath beds of rock-salt of marine origin (sometimes replaced by anhydrite) and always separating from sea-water, which undergoes concentration before the rock-salt separates out.

Such rock gypsum is commonly massive and sometimes impure, but pure white gypsum (alabaster) also occurs. It also occurs as crystals (selenite), scattered through mud and sand deposits, especially in desert regions. A fibrous variety (satin spar) occurs in veins. Gypsum is seldom formed in extensive beds in desert regions except where sea-water evaporates. It is also formed as an alteration product, generally of limestones, by waters carrying sulphuric acid; large beds of *gypsum* result in this way, those of New York state being an example.

Anhydrite deposits are also changed to gypsum when surface waters come in contact with them. Extensive deposits of gypsum are found in the"Red Beds" of the western United States, and in Kansas, Oklahoma, and Texas. It is found in Nova Scotia and in the Paris Basin, where it has been and still is extensively quarried and burned into Plaster of Paris; also in many other parts of the world.

Raw gypsum is ground and used as a natural fertilizer (land plaster), to retard the setting of cement, and for many chemical purposes. When burned or "calcined" at 350° F. it loses most of its water ($CaSO_4$ 1/2 H_2O), and is ground into the familiar "Plaster of Paris." When this is mixed with water gypsum is again formed. It hardens rapidly and is used extensively in the arts for molding, statuary, etc., and for stucco work.

Anhydrite

Calcium Sulphate ($CaSO_4$)

This is distinguished from gypsum by its greater hardness and specific gravity. Its colour is often also more grayish. It is formed as a primary deposit from sea-water in cut-off basins, especially when the water contains an excess of chlorides.

The largest known deposits thus formed are in northern Germany, where they underlie the salt and potash beds. Anhydrite slowly changes to gypsum by taking on water, with an expansion of the mass. It is of little economic importance.

Carbonate of Lime

Calcite, Aragonite, Limestones, etc. ($CaCO_3$)

The great bulk of the deposits of carbonate of lime, including the limestone beds, is of organic or of elastic origin, but there are a number of lime deposits which belong to the hydrogenic division. The most extensive of these is probably calcareous tufa, which is formed by springs issuing in limestone regions and depositing the excess of lime which they hold in solution.

Calcareous tufa is mostly porous and light, incrusting not infrequently mosses and other plants and also other objects. Some massive deposits of this type, however, are known, constituting the so-called "Mexican Onyx." Compact carbonate of lime deposits are also formed in caves as stalactites and stalagmites.

These usually have a banded structure, showing the successive addition of layers. That of the stalactites is concentric around the longitudinal axis, which in the early stages is formed by a delicate tube. Beds of limestone are sometimes built up from large rounded masses or "concretions," which have resuited from the deposition of carbonate of lime, generally around a nucleus.

Sometimes these are formed as original masses of limestone, as in the limestone deposits of Lake Lahonton or an older limestone of this type which occurs in the Permian series (Magnesian limestone) of northeastern England. In other cases these concretions occur in shale beds, forming distinct layers, which when they become confluent by continued growth form bands of limestone.

When separate they often take on disk-shaped or spherical forms, while a series of radial cracks are developed in the interior with the growth of calcite veins. Such concretions are called *septaria*. All carbonate of lime deposits, effervesce readily with dilute hydrochloric acid, this being the most distinctive test.

Dolomite

Carbonate of lime and magnesia ($(CaMg)CO_3$) — This differs from the pure carbonate of lime deposits in its greater hardness and in the fact that it effervesces only in strong hydrochloric acid. It is sometimes a primary deposit in basins where anhydrite is forming, with which it becomes more or less intimately mixed.

It may also be formed as an original precipitate in pottions of the sea in which the solution has become concentrated. Many dolomitic limestones are, however, of secondary origin, the original magnesia content, which was derived for the most part from calcareous algae, etc., being proportionately increased through the solution of carbonate of lime by ground water.

Secondary deposition of magnesium carbonate also takes place from solutions in the circulating ground water. Pure dolomite contains 54-35% $CaCO_3$, and 45.65% $MgCO_3$.

Magnesite

Carbonate of Magnesia ($MgCO_3$)

This is harder and more compact than dolomite, and dissolves with effervescence only in hot hydrochloric acid. It occurs as a crystalline mineral, as replacement of dolomite, and also as an amorphous, earthy, hard, compact mineral, probably a colloidal precipitate.

It is often concretionary with conchoidal fracture, appearing like unglazed porcelain, this type being usually derived from the alteration of serpentine and other magnesian rocks. Beds of magnesite are found associated with gypsum in fresh-water limestones of France.

Apatite and Phosphate Rock

Although the most extensive deposits of phosphate of lime are of organic origin, there are some that must be considered as purely aqueous or hydrogenic deposits. Among these are the apatite veins and the marine concretions of phosphate around an organic nucleus. Phosphate of lime occurs also as a secondary replacement of limestone.

Apatite crystallizes in hexagonal prisms, etc., and is readily recognized by its form, hardness, and colour. Rock phosphates are generally amorphous and compact. The pure mineral (tricalcium phos phate) contains 45.8% of phosphoric acid (P_2O_5). The principal use of phosphate is as a fertilizer.

Potash Salts

There are a number of evaporation products which are primarily salts of potash and are important sources of this substance. Among the more common of these are *Sylvite,* the chloride of potassium (KCI), a very soluble, soft, transparent, milky reddish or yellowish mineral, commonly mixed with rock salt; *Carnallite,* the hydrous double chloride of potassium and magnesium ($KCI \cdot MgCl_2 \cdot 6\ H_2O$), a colorless or snow-white or variously colored salt, easily soluble, and an important source of potash; and *Kainite,* the compound chloride of potassium and sulphate of magnesium, with water ($KCI \cdot MgSO_4 \cdot 3\ H_2O$), colorless to deep blood-red, and chiefly an alteration product. The chief use of potash is for agricultural purposes and in chemical works.

Trona

Sodium Carbonate ($Na_2COs \cdot NaHCO_3 \cdot 2\ H_2O$)

This is a common evaporation product of alkaline lakes. It is glassy or transparent when crystallized, but usually forms a white salt.

Important American localities are Searle's Marsh, Owens and Mono Lakes, Cal., and Soda Lake, Nevada, while some Russian and Hungarian lakes, and the Natron lakes of Egypt west of Cairo, represent foreign localities. It is used for domestic and chemical purposes, but much of the commercial trona is artificially produced. On account of its solubility it is not generally found in the older rocks.

Mirabilite

Hydrous Sulphate of Sodium ($Na_2SO_4 \cdot 10\ H_2O$)

This is a crystalline, granular salt, usually colorless, and readily soluble in water. It is formed, especially in winter, by certain saline bodies, such as Great Salt Lake, Utah, and the Kara Bugas Gulf on the east coast of the Caspian Sea.

On the floor of the Kara Bugas a bed of mirabilite, estimated to contain 1000 million metric tons, has formed but it is not yet exploited. In Wyoming and New Mexico occur beds of this salt mixed with epsomite, natron, and common salt (halite). In some cases the deposit is 15 feet thick and covers an area of 100 acres.

Glauberite

Double sulphate of sodium and calcium ($Na_2SO_4 \cdot CaSO_4$). -This is a white, gray, yellow, or red mineral, a little harder than common salt (2.5-3). It occurs in many playa lakes (Borax Lake, Searle's Marsh, Death Valley), and in the salt deposits of Germany, Spain, Austria, Sicily, etc. It also occurs in a number of Tertiary deposits in Spain.

Soda Niter, Chile Saltpeter

Nitrate of Sodium ($NaNO_3$)

This is a white to reddish brown, gray, or lemon-yellow salt, commonly impure, when it is called *caliche*. It is abundant in the desert tracts of western Chile and elsewhere in South America and other parts of the world.

Borax Salts

These include *Borax* or *Tinkal*, sodium borate ($NaB_4O_7 \cdot 10\ H_2O$), a colorless or yellowish or green to gray, soluble mineral, somewhat harder than common salt; *Colemanite*, the hydrous borate of calcium ($CaB_4O_{11} \cdot 5\ H_2O$), generally a massive, glassy or colorless, more or less transparent mineral, found as a bed from five to 20 feet thick in San Bernardino Co., Cal., and elsewhere; and *Ulexite*, the double borate of sodium and calcium with water ($NaCaB_5O_9 \cdot 8\ H_2O$), which occurs as white balls of fibrous material and satiny luster (cotton balls).

Borax salts are found in volcanic regions, such as the famous "saffioni," *i.e.*, fumaroles, in the volcanic region of Tuscany; in hot spring and lake deposits of volcanic districts, as in Tibet, etc., and the Coast Ranges of California, and in playa deposits, as in Death Valley and elsewhere in the western United States. Colemanite is the chief American source of borax.

Silica

Oxide of Silicon (SiO_2) Frequently with Water

The most familiar form of this is quartz in crystalline or amorphous form, which occurs in veins, geodes, etc., colorless to variously colored; also as addition to quartz grains in sandstones, enlarging them to fill all the interstices. It occurs extensively as a porous, white, and light spongy deposit of hydrous quartz, around geysers and is hence known as *geyserite* or silicious sinter.

Again, it occurs as concretions of *flint* in chalk beds, or as chert nodules or layers in limestone. These types are compact and have a conchoidal fracture, with a black or dark brown colour on the freshly fractured surface. Flint and chert are secondary concentrations through solution and redeposition of disseminated silica particles of organic origin.

Glauconite

Greensand

This is a complex silicate of potassium and iron ($K{\cdot}FeSi_2O_6 \cdot H_2O$). It consists mainly of dark green grains, generally mixed with impurities, found in strata of Cretaceous and other formations, and is forming at the present time along the margin of the continental shelf in the Atlantic Ocean and elsewhere. It promises to be an important source of potassium, in which the New Jersey deposits are especially rich. They also contain much phosphorus.

METALLIC AQUEOUS DEPOSITS

Manganese Ores

Nodules of oxide of manganese occur in many parts of the deep sea in the region of red clay deposits. They are more or less rounded or irregular masses of black colour. Oölitic pyrolusite (MnO_2), an iron-black mineral of metallic luster, occurs interstratified with other beds in certain localities. A hydrous oxide of manganese occurs as an amorphous earthy deposit in bogs, generally associated with iron ores.

Limonite

Bog iron ore, Hydrous Oxide of Iron ($2Fe_2O_3 \cdot 3H_2O$)

This occurs in massive, botryoidal, earthy, or porous masses of yellow or ochery colour and yellow streak. It forms deposits in bogs and swamps, in shallow water, in depths above 12 feet, from the oxidation of carbonate of iron carried into the bog in solution. The ores are always mixed with sand or earthy impurities. There are several other ferric hydrates, some with less and some with more water; *e.g.* Göthite ($Fe_2O_3 . 2H_2O$). Oölitic limonites occur in older strata, where they form an important source of iron, though of low percentage. Such are the *minette* ores of Lorraine and Luxemburg, formerly one of the chief domestic sources of iron in Germany. These are believed by many to have resulted from the oxidation of oölitic siderite or glauconite.

Hematite

Red Oxide of Iron (Fe_2O_3)

This ranges in colour from deep red to reddish brown and black with red streak. It occurs interbedded with shales and limestones in the Palæozoic formations of Germany, France, Bohemia, and the United States, where the iron ores of New York and the Appalachian region form a characteristic example. Many of these beds appear to be replacement of limestone by the iron; in other cases the iron ore seems to be a primary deposit.

Siderite

Carbonate of Iron ($FeCO_3$)

This important iron ore occurs in crystallized form in veins cutting limestone and other rocks, and is readily recognized by its perfect rhombohedral cleavage, vitreous to pearly luster, greenish to brownish colour, and translucent to subtranslucent character.

It is more common, however, as an interbedded rock in the older sedimentary series, being known as clay-iron-stone, spherosiderite, or black band. Clay-iron-stone has a dense or fine-grained structure, forming concretions, which often include organic remains. The "black band" forms continuous layers in the formations which carry coal.

ORIGIN OF THE AQUEOUS OR HYDROGENIC ROCKS

Each of the several water bodies of the earth may form precipitates of mineral matter, and hence we may classify these deposits under the following heads:

- *Marine* deposits, or those formed in the sea and its dependent water bodies.
- *Lacustrine* deposits, or those formed in lakes, ponds, freshwater marshes, salinas, playas, etc.
- *Fluviatile* deposits, or those formed by rivers in their beds, or on the flood-plain or delta surfaces, except those formed in lakes along the river course, or at its mouth.
- *Terrestrial* deposits, or those formed by springs and by the ground water in fissures, caverns, cavities in the rock, etc. To these belong many important deposits of mineral matter. The several types may grade one into the other, but their characteristics are quite distinct.

SEA-WATER AND THE EVAPORATION PRODUCTS

Amount of Salt in Sea-Water

The preceding the oceans, which are the large bodies of sea-water lying between the continents, contain the normal salt water, in which about 35 grams of salt occur in every liter of water. Since a liter of pure water weighs 1000 grams, the quantity of salt is essentially 35 per thousand by weight, which is expressed by the formula 35 per mille (or 354 0/00).

This corresponds, of course, to 3-5 per hundred or 3.5 per cent (3.5%), but since the difference in salinity between different water bodies is often very slight, and because in brackish and fresh waters the actual quantity of mineral matter in solution is very small, it is more satisfactory to express the quantity in permillages, than in percentages.

The quantity of salt in solution determines the *salinity* of the water. Thus the average salinity of the ocean water is 35 per mille, (3.5 per cent), varying somewhat for the different oceans, for different parts of the same ocean, and for different depths. On the other hand, the salinity of the Red Sea surface waters is 38.8 per mille, while that of the surface waters of the Black Sea is only 18.3 per mille, which is due to the fact that this water body is almost entirely cut off from the rest of the sea, and that it receives many freshwater rivers.

Finally, the average surface salinity of the Baltic Sea is only 7.8 per mille, whereas if the water of this enclosed basin is taken as a whole, it is somewhat more saline because of the greater salinity of the deeper layers. Even then, however, it is only 10 per mille. The Baltic Sea, moreover, shows a remarkable gradation in the salinity of its waters from west to east.

Where it joins the North Sea at the Skager Rack, the salinity is 34 per mille, but in the Kattegat it is only 22 per mille. Thence it gradually decreases eastward and northward until the waters near the heads of the Gulf of Finland and that of Bothnia are essentially fresh.

COMPOSITION OF THE SEA-SALTS

Although, strictly speaking, the material held in solution by the water of the sea is not in combination as salts, such as are produced on evaporation, but rather in the form of ions, the basic elements and the acid radicals being separated, nevertheless it is customary and convenient to consider them as combined into the form of salts.

Among these, common salt or sodium chloride makes up the bulk of the material, being nearly 78 per cent of the total mass of salt, or over 27 per mille of the salinity (which is taken as 35 in round numbers). In the following table the composition of the sea-water salts is given in the form of such combinations, together with the permillage of each in normal sea-water, and the number of short tons in a cubic mile of sea-water.

With the calcium carbonate are included the small quantities of other salts present, such as the iodine, lithium, manganese, and phosphorus salts, and the silver, gold, nickel, and other metals which are present in minute quantities in the solution.

SALTS PRODUCED BY EVAPORATION OF SEA-WATER

The two principal salts which are produced by the evaporation of sea-water are the common salt, sodium chloride ($NaCl$), and gypsum or calcium sulphate, with two molecules of water ($CaSO_4 + 2\,H_2O$). By local oversaturation with lime, calcium carbonate ($CaCO_3$) may also separate out, but this is more commonly produced by chemical reaction. When the evaporation has gone very far, other salts, such as those of magnesium, and finally potash salts, separate out. Common salt and gypsum are obtained on many sea-coasts, either by evaporation of the water by artificial heat, or by conducting the sea-water into large shallow "pans," that is, fields surrounded with dams and having a hard, flat bottom.

When the outlet of the pan is closed, evaporation takes place under the influence of the sun and drying winds, and after a while gypsum separates out. When most of this has been deposited, the water is conducted to another pan, where evaporation continues until a large part of the common salt (sodium chloride) has separated out.

Then the remaining dense brine, which is called the *mother liquor*, is drawn off and either returned to the sea or further evaporated for the rarer salts. In this manner a large part of the world is supplied with its domestic requirements of salt, though vast quantities of this commodity are also obtained from inland salt deposits, which are the product of evaporation either of sea-water or of inland salt-bearing waters, in former geological periods.

EXPERIMENTS IN EVAPORATION OF SEA-WATER

In 1849, the Italian chemist, J. Usiglio, published the results of experiments which he had made at Cette on the south coast of France. He had taken 5 liters of the sea-water from the Mediterranean and evaporated this, keeping an exact record of the point which the evaporation had reached when separation of the several salts took place, and determining the amount of the various salts separated at the successive stages.

No separation of salts occurred until the water was evaporated to nearly one half its volume, when the iron oxide and a part of the carbonate of lime of the sea-water separated out. Later still, when the original 5 liters of the water had been reduced by evaporation to about one liter, the remainder of the carbonate of lime, together with the hydrous sulphate of lime, or *gypsum*, was precipitated.

More than 84 per cent of the total amount of gypsum contained in the sea-water was deposited before the water became dense enough to allow separation of the common salt (NaCl), the remainder of the gypsum being thrown down with that salt and with various amounts of magnesium salts ($MgSO_4$ and $MgCl_2$), and finally with sodium bromide. A significant fact was that no sodium separated out until the evaporation had reduced the original 5 liters to less than half a liter, or to less than one tenth the original volume.

At that time the amount of solid matter still held in solution in the half liter of water remaining was about 184.4 grams, which would correspond to 368.8 grams per liter, or a salinity of 368.8 per mille, whereas the salinity of the original sea-water was about 38.5 per mille.

From this we must conclude that a water body must reach this high degree of salinity by evaporation before salt can be deposited in nature. Since there are to-day no known large bodies of water with such a salinity, it follows that extensive salt deposition is not going on to-day by simple evaporation of large bodies of sea-water.

To be sure, there are many small and shallow marginal lagoons on the sea-coast, and especially on the shores of more or less enclosed salt-water bodies, such as the Black and Caspian Seas, and Great Salt Lake, where evaporation goes far enough to precipitate salt — but this is, as a rule, only in comparatively small amounts, though commercially important.

Complete evaporation of the 5 liters of sea water was not achieved by Usiglio, for his experiments ceased when the volume had been reduced to about 81 cubic centimeters. This remaining dense "mother liquor" retained all of the potash salt of the original sea-water in solution, together with some of the sodium and magnesium salts.

The salts of the mother liquor are precipitated only at very high or very low temperatures, and from this we must conclude that potash deposits in nature are formed only under exceptional conditions.

SPECIAL CONDITIONS FAVORING DEPOSITION OF SEA-SALTS

From the foregoing it becomes apparent that the first requisite for the deposition of sea-salts in nature is the concentration of the sea-water under the influence of conditions which favour evaporation, such as the heat of the sun and, above all, drying winds. That such evaporation cannot go on in the open ocean is apparent, and so we must look to bodies of sea-water cut off from the main oceans.

The Caspian Sea

This, the largest isolated saltwater body of the earth, may in many respects be considered typical. It lies within the region of drying winds, and is partly surrounded by deserts. Evaporation has gone so far that its surface is 85 feet below sea-level, yet the salinity of its water is only 12.94 per mille, or a little over one third that of normal ocean water. This is due to the fact that a large amount of fresh water is brought in by the Volga and other rivers tributary to it, so that in spite of the evaporation, the salt content is very low.

The original salt has been specially concentrated in the Kara Bugas Gulf and other dependent bodies, and some salt has no doubt been deposited on the bottom of the lake and then preserved by a covering layer of impervious material (gypsum, clay, etc.).

The Black Sea

This nearly, but not quite, isolated mediterranean water body has a surface salinity of only 18.3 per mille, but the lower layers are denser, so that the average salinity of the water as a whole is 22.04 per mille. Obviously no salt can be deposited in the open parts of this sea.

There can be no question, however, that were it not for the supply of fresh water, both the Black and the Caspian seas would have a high salinity, while the Baltic would be nearer to sea-water in that respect. Indeed, should the fresh-water supply be entirely cut off from the Caspian, continued evaporation would result in the separation of most of its salt upon the bottom of its basin, and eventu ally only a layer of mother liquor would remain to cover these deposits, and under special conditions this mother liquor might also be forced to part with its salts.

That such evaporation of large enclosed bodies of sea-water has occurred in the past history of the earth is indicated by the nature of the salt deposits now found enclosed in the rocks of the earth's crust, as will be seen presently. To-day such deposits are formed in local "salt pans" along the margin of the Black Sea, the Red Sea, and especially in the Ran of Cutch on the west coast of India, deposits which for centuries have supplied the natives with salt.

Effects of Condensation of Sea-Water on Its Animal Life

A moment's reflection will show that in the process of evaporation and concentration of the cut-off portion of the sea-water, all the animals which lived in that water would be killed, and their remains would sink to the bottom or be cast upon the shores. The shells of the Mollusca, the horny coverings of Crustacea, and the bones and teeth of fish and other vertebrates would be embedded in the layer upon which the salt later comes to lie.

Thus a very definite and restricted fossiliferous substratum is produced for salt deposits of this type, and this will furnish a criterion by which ancient salt deposits can be interpreted. If the change in salinity is gradual, because the water body subject to evaporation is large, extensive fossiliferous deposits may be formed, including important beds of limestone, before the water is dense enough to kill the organisms.

After that the water will remain essentially lifeless (though there are certain forms of animals which live only in strong brines), and the deposits formed in it will be barren of organic remains. An exception to this may, however, occur if the sea should break into the basin again, flooding it with normal seawater, and bringing in with it the normal sea-fauna. Then, if the basin is again cut off from the sea, evaporation will set in with the repetition of the series of evaporation deposits.

The Deposition of Sediments and Salts

If the basin whose waters become subject to evaporation is large, the waters, as they shrink, will leave a succession of deposits of various kinds around the margin. Along the shores would be found sands and clays which farther out might merge into organic limestone. As the sea-water becomes concentrated, the organically formed limestones would come to an end, and chemically formed limestones and dolomites would take their place. Gypsum or anhydrite deposits follow next, and finally, as the water becomes very saline and the area much contracted, salt is deposited. Last of all, the mother liquor salts are precipitated under favorable conditions, but only in the central area of the original basin, where the last of the water lingers. It must be emphasized that normal salt deposits formed from evaporating sea-water must always be underlain by a layer of gypsum or its anhydrous equivalent, the mineral anhydrite.

ROCK-SALT DEPOSIT FORMED BY EVAPORATION OF SEA-WATER

Among the many salt deposits within the earth's crust which were formed during earlier geological periods and preserved through burial by later deposits, a certain number can best be explained as formed by the evaporation of cut-off portions of the sea in the manner above outlined. This means, of course, that they have the essential characteristics which we have seen are the normal accompaniments of salt deposits formed from complete evaporation of a cut-off body of sea-water.

The most notable example is found in the great salt deposits of North Germany most widely known as the Stassfurt deposits, though this is only one of the localities where these salts are mined. Their peculiar interest lies in the fact that they have associated with them the rare potash salts which are precipitated from the mother liquor on complete evaporation of the sea-water.

Before the World War these deposits furnished by far the largest amount of potash to the commerce of the world, and their abundance is such that they can supply the entire world at the present rate of consumption for perhaps 2000 years to come.

If we take a section through these deposits from top to bottom, as revealed by the numerous boreholes and by mines in operation, we find the following characteristic succession:

At the base lies a variable thickness of limestones and dolomites, which in some sections contain old reefs formed by organisms which inhabited these waters while they were still connected with the sea, and for some time after. This limestone is known as the *Zechstein*, and it takes its definite place with the salt deposits and with the underlying red sandstones, in the series of successive formations which were made during the later portion of the Palæozoic era of the earth's history.

Above the Zechstein limestones and dolomites lies a formation of anhydrite and gypsum 100 meters in thickness, and this is followed by the salt beds. These average 245 meters in thickness and are subdivided by about 3000 layers of anhydrite, the so-called annual rings. Above this follow the mother liquor salts which have given these deposits their great value. They include more than 30 rare minerals, the more abundant, as a rule, in layers or strata, the whole averaging from about 60 to over go meters in thickness.

The most important of them from a commercial point of view are, of course, the potash salts. Above these mother liquor salts occurs a second series, beginning generally with a salt clay containing remains of marine animals, followed by anhydrite (30-80 meters) and rock-salt, with about 400 annual rings of *polykalite,* which is a complex hydrous sulphate of calcium, magnesium, and potash. The series is closed by a succession of minor layers of red clay, anhydrite, rock-salt (about 40 meters), anhydrite, and red clay, the last forming the top of the deposits.

Throughout the entire series, except in the basal Zechstein limestones and the salt clay at the middle, remains of organisms are wanting, except a few plant fragments, which were blown into the water body. It appears then that both series of deposits, the lower as well as the upper, represent the succession which we have seen is characteristic of a progressively drying up cut-off portion of the sea. Moreover they differ from the succession of salt deposits formed in other ways, and they may, therefore, unhesitatingly be interpreted as accumulations of the first type.

The regular intercalation of the anhydrite layers in the lower salt suggests that they are due to seasonal fluctuations in the density of the water — a periodic slight reduction in the salinity during a moister period putting a temporary end to salt deposition and permitting the formation of anhydrite, which is separated from less concentrated waters rich in salt.

If these changes were of annual recurrence, as seems likely, it would appear that the formation of this salt mass took three thousand years. Each annual ring is on the average about 7 millimeters thick, while the salt layers with which they alternate are from 8 to 9 millimeters in thickness.

After the deposition of most of the mother liquor salts, a period of fluctuation occurred, and salt with polyhalite layers was formed. This alternation, too, is probably seasonal, but shows that the brine was now more highly concentrated and of changed composition, and it is probable that regular changes in temperature were most influential in producing the succession of deposits.

The higher anhydrite and salt series indicates that the sea-water again filled the basin, then it was cut off anew, and again underwent complete evaporation. It is also quite probable that the waters of these basins were enriched by salt brought in solution by intermittent streams from the surrounding hills, as is the case in many modern desert basins, and that, because of this excess of salt, anhydrite, rather than gypsum, was deposited.

We cannot, however, settle the question of origin from the deposits of North Germany alone. If they were formed by the drying up of an enclosed sea, there should be neighboring salts and sediments which should show such variations as we would expect nearer the shores of this water body, and these deposits should be of the same age.

The method of searching out and determining formations of like geological age in different parts of the world will be dealt with later. Suffice it to say here that such age determination and correlation of formations in different countries is quite possible by the aid of fossils, by the relationship of the beds to one another, and by other criteria.

Taking, then, such deposits of the same age in other parts of Europe, we find first that the mother liquor salts fail as we proceed away from the North German region, where apparently was the centre of concentration.

Salt is still found in a number of sections even in eastern England. Here, however, magnesian limestones prevail, some of them showing a peculiar structure suggestive of chemical deposition. Limestone formed of a restricted number of organic remains elsewhere makes up the deposit of Permian age in England.

Finally, in the north and west of England, only red sandstone deposits represent the Permian, these being formed near the old shoreline of the Permian basin, and in part above it, by rivers which brought sands from the uplands and dropped them in their lower shallow courses.

In the other direction, toward the Ural Mountains of Russia, a change in the character of the Permian formations may also be observed. At first limestones predominate, but farther east these are largely replaced by sandy and clayey beds, among which coal seams are found, which indicate a swamp-land condition.

It seems practically demonstrated, then, that these ancient salt deposits with their valuable beds of potash salts were formed by the drying up of a large [body of sea-water which had become separated from the main ocean by the formation of a land barrier. This water body appears to have extended from central England on the west nearly to the Ural Mountains on the east. Its southern boundary was north of the Danube, and its northern in the region of the Baltic Sea of to-day. A significant corollary is that the climate of North Europe was very much dryer at that time than it is to-day, for at present no such complete evaporation would take place. This will be referred to again in a later part of this book.

CONCENTRATION IN LAGOONS BAR THEORY OF OCHSENIUS

Complete evaporation of a salt lake is only one way — though a most effective way — of producing salt deposits. A second method consists in the concentration of the sea-water in a nearly shut-off lagoon in regions of and climate. As a typical example of this we may select the Kara Bugas Gulf a bay which lies on the eastern coast of the Caspian Sea, from which it is separated only by a sand bar, across which a narrow strait maintains connection with the Caspian.

On the other sides the gulf is surrounded by deserts, and there are no streams entering it. Since the Caspian Sea is a large salt-water body, though of lower salinity than the ocean, the Kara Bugas Gulf is nearly as satisfactory an illustration as a similar bay on the open sea would be.

Salt Deposit Formed by Bar Theory

Perhaps the best example of an older rock-salt deposit which can be explained by the Bar Theory was found in the salt beds of the Bitter Seas on the peninsula of Suez. Before the Suez Canal was cut, these lakes were of very high salinity, and on the bottom of at least one of them an immense bed of rock-salt had formed. This has, however, been entirely redissolved by the fresher waters of the canal.

A characteristic feature of this salt bed was not only the presence of many layers of gypsum, but also of innumerable layers of clay in which were embedded the shells of mollusks such as now live in the Red Sea near by, while similar shells were entombed in the sediments which formed on the bottom of the Red Sea during the period of salt deposition, though these sediments are free from salt.

We have here a deposit which satisfies all the requirements of the lagoon and bar example of the Kara Bugas, namely, a circumscribed area, presence of numerous organic remains in the saltbearing series, and the association in a neighboring area (the Red Sea) of normal deposits with the same fossils but with no salt. In the present case it is known that these Bitter Lakes formerly constituted an extension of the Gulf of Suez and the Red Sea, forming the ancient Heroöpolitan Gulf.

The silting up of the mouth of this gulf, which was not yet complete during the sixth century before Christ, formed the bar which cut off the lagoon from the remainder of the gulf. According to some authorities, this bar appears to have been the site of the crossing of the Red Sea by the Israelites in their exodus from Egypt.

With the formation of this bar, which now constitutes the southern margin of the isthmus of Suez, the conditions favorable to the deposition of the gypsum and salt were produced. Repeated overflow from the Red Sea supplied the waters and the mud in which were buried the organisms which could live here until the water became too salty.

Then their shells and other remains sank to the bottom; a layer of gypsum was formed over them, and then a layer of salt, covered by a more or less impervious layer which prevented re-solution when next the waters of the Red Sea poured in again. In the waters thus freshened, new organisms developed from the larval stages brought by these waters, and a new cycle of deposition was inaugurated. The mother-liquor, however, never evaporated in this lagoon, but remained behind, forming the bitter waters of the lakes, which, however, have lost much of their character since the letting in of the sea-water by the canal.

DEPOSITION OF SALT IN INLAND DESERT BASINS

Salt deposits are forming in many portions of the world to-day, where no direct connection with the sea exists. The inland salt lakes and salinas are in some cases shrunken bodies of fresh water formerly of greater extent. This is the history of Great Salt Lake of Utah, which is only a remnant of a much larger fresh-water lake.

Lake Bonneville. By evaporation and concentration of the water in the deeper part of the basin, the salinity was increased, and finally in some cases reached the concentration necessary for the deposition of salt. Much additional salt is constantly brought into the lake by the streams which feed it. Naturally the question as to the origin of the salt in these waters arises. The answer is that it is leached out of the rocks within the drainage area of the basin.

Not all rocks contain sodium chloride, but this is generally present in clastic and other rocks which have been formed on the bottom of the sea in former geological periods. These rocks include ancient sea-water, often highly concentrated, within their pores, where it is hermetically sealed up and is set free only when the rocks are subsequently exposed to atmospheric decay and erosion.

Then the *connate water,* as it is called, evaporates, but the salt remains behind to be redissolved by the surface waters and carried away. If it is carried into an inland drainage basin from which there is no escape, except by evaporation, salinas and even extensive salt deposits are formed. Such areas, covered with glistening white rocksalt or with irregular salt masses, are found in the desert basins of central Asia, and indeed are not uncommon in many other desert areas.

So long as the supply of salt lasts, such a deposit will continue to grow by periodic additions, and thus beds of salt of great thickness may form. If silt is brought in during a period of flooding, or if sand or dust is blown across the salt plain, a layer of clastic sediment may form over the salt, and this in turn may again be covered by pure salt deposits.

As a rule, however, gypsum is not deposited in such basins, not because the connate waters imprisoned in the rock did not contain it, but because after it is set free it is less readily redissolved by the surface waters than is the common salt, and that which, is dissolved is likely to separate out again from the waters before they reach the central basin. In conformity with this we often find the sands which surround such basins filled with gypsum crystals which have grown from the ground water as it passed through the sands, carrying the more soluble sodium chloride to the central basin.

In like manner the potash salts will not, as a rule, reach the central basin, for though they are very soluble, the fine particles of the soil of the desert among which the water must find its way to the central basin have a great affinity for the potash and will, by some still little understood process, separate this substance from the water during its passage. Hence the water which reaches the central basin will contain little else than pure sodium chloride, and therefore only pure rock-salt deposits are formed.

Gypsum may, however, result from the alteration of limestones formed of lime-mud and dust which is washed or blown over the salt. In this manner gypsum beds may be formed above the salt beds, whereas in normal marine deposits of salt, or in those due to the evaporation of cut offs, the gypsum will underlie the salt.

The importance of these facts in the determination of the origin of older salt beds is very great. Another fact that must not be overlooked in desert salt deposits is that the parting and enclosing layers of sediment will contain no marine organisms. They will, indeed, be for the most part entirely free from organic remains. A few desert organisms and migratory birds may, however, become entombed in these deposits or even in the salt itself, but their terrestrial character is readily recognized by the expert.

Huntington reports finding, in the salt of Lop Nor in eastern Turkestan, a dead plover which had been preserved in the salt for centuries. Around the border of some of the Persian salinas a zone of mud is sometimes found, in which are entombed the bones of animals which came to drink of the salty water and perished there.

Beyond these salinas in all directions we pass into the region of desert sands and dust deposits, and these have very definite characteristics which can be recognized even after they have hardened into rock beneath a cover of other deposits. Thus the geologist will generally be able to recognize in the rocks associated with the ancient desert salts the structures which clearly indicate that origin. The appearance of a typical American salt-marsh or playa.

Among the geologists who have made extensive investigations into the origin and mode of deposition of desert salts, the foremost rank must be assigned to Professor Johannes Walther of the University of Halle.

An Ancient Example of a Desert Salt Deposit

In the central part of the state of New York, in western Ontario, and in southern Michigan, occur ancient rock salt deposits, all of which were formerly and some of which are still buried under thou sands of feet of rock, a considerable portion of the latter being of marine origin. The rock-salts themselves, however, which rest in some places upon marine beds of Lower Silurian (Niagaran) age and in others upon non-marine sediments, are in some places covered by marine beds of Upper Silurian age (Monroan) and are associated laterally with deposits of clastic material which shows all of the features of sediments found in modern deserts.

Moreover, no fossils are found in these deposits nor in those which separate the several salt beds of the series from one another, nor are there any Middle Silurian beds of marine origin to be found within hundreds of miles of the salt deposits. Thus it appears that these very ancient salt beds of Middle Silurian age were formed in a desert which then occupied much of the area now covered by the Great Lakes and adjoining territory.

No potash deposits are found associated with these salts, and from what he has learned so far, the student will realise that there is little likelihood of the finding of these salts unless it can be shown that the basal salt beds of some sections are not of desert but of marine origin, resulting from the drying up of the last remnant of the Niagaran sea which preceded the desert period.

There is, however, one important fact which seems to argue against such a marine interpretation of the basal beds, and that is the absence of gypsum or anhydrite beds beneath the salt. Indeed, the entire series of Salina salt deposits lacks the foundation layers of gypsum or anhydrite, though gypsum overlies the salt in a number of localities. Much of this is, however, known to have been produced by the alteration of former limestone beds which were invaded by sulphur-bearing waters.

CARBONATE OF LIME DEPOSITS

Although some carbonate of lime separates out on evaporation of the sea-water, this is of such small amount that it practically disappears in the evaporation series produced along the seacoast and from cut-off bodies. Nevertheless, carbonate of lime deposits are formed in the sea, not so much by evaporation, — though such an origin may be ascribed to some deposits, — as by the force of attraction of other particles of lime in sea-water saturated with lime carbonate, or by the abstraction of the solvent carbon dioxide through agitation of the water.

Chemical precipitation of lime also takes place and is perhaps the most common mode of lime deposition in some places aside from that due to organic action. This precipitation, however, is due largely to the formation of ammonium carbonate by the decay of organic matter, and this ammonium carbonate reacts with the lime sulphate or other lime salts in the sea-water, producing calcium carbonate, which is precipitated, and an ammonium salt which remains in solution. The reactions may be written in the following way:

$$\underset{\text{Calcium Sulphate}}{CaSO_4} + \underset{\text{Ammonium Carbonate}}{(NH_4)_2CO_2} = \underset{\text{Calcium Carbonate}}{CaCO_2} + \underset{\text{Ammonium Sulphate}}{(NH_4)_2SO_4}$$

Or again,

Where lime is precipitated by such reactions, it often forms more or less spherical or irregular masses or nodules, to which the name *concretions* is applied. Such concretions have been dredged from many portions of the sea, and they appear to be especially common where organic matter which has reached the floor of the ocean undergoes a process of decay.

Such areas are, however, not universal because the organic matter on the ocean bottom is generally devoured by bottom-feeding animals before the decay progresses far. It is only where the character of the water, or the temperature, is such that bottom-feeders are scarce or absent, that decay of stray organic matter can take place.

Deposition of lime due to the attraction of other lime particles is illustrated by the hardening on the ocean bottom of the loose limesands and muds worn from the coral and other organic limestone masses in the sea. It is a general fact that wherever lime-mud, or sand, forms upon the sea-floor, this is soon bound together by the filling in, between the particles, of lime derived from the sea-water. This seems to take place most actively in warm regions, where the amount of lime in the sea-water is above the average. As a result, the floor of the ocean in such regions is a hard surface to which various stationary marine animals attach themselves, while others, such as certain worms or sponges, bore into this rock to a certain depth.

On the surface of ancient limestone beds we often find the marks of animals which had become cemented to it. Such cemen tation could of course take place only if the surface were of sufficient firmness, and this indicates that the old deposits of lime-sand, or mud, from which these beds were formed in the sea, hardened by further separation of lime, so that the animals living there could become attached to it.

Another illustration of such lime deposition is seen in the coating of lime around grains of quartz, basalt, or other sand, or around fragments of shell, etc., which are found on the shores of the Island of Gran Canaria in the Canary Islands. These coated grains are more or less spherical and of the texture called oölitic. By further deposition of lime between the grains they are bound together into a solid rock, an oölite, which is quarried at low tide.

The water here is warm, having throughout much of the year a temperature above 20° C., and there is much lime in solution. Still another interesting example of lime deposition is seen in certain Mexican lagoons where insect eggs are coated with lime, producing a series of rounded oölite grains. Most of the grains of oölitic character are, however, produced by the activities of bacteria or minute algæ in the sea, and on this account must be classified as of organic origin.

OTHER CHEMICAL DEPOSITS IN THE SEA

Small quantities of other substances are deposited in the sea as the result of certain chemical reactions, or through attraction by material of like composition. The most important of these are phosphatic concretions and concretions of oxide of manganese, both of which have a wide distribution over the sea-bottom. A third group of such deposits forms grains of the green mineral glauconite, of which "green-sands" are made.

Phosphate of Lime

This is produced by many marine animals which take the phosphoric acid either directly from the sea-water, or from their food. The phosphate of lime is built into certain hard tissues, as the shells of some brachiopoda the bones and teeth of fish, etc., and it is also present in the excrements of fish and other animals. Such particles accumulating on the seabottom have the power of attracting to them more phosphate and precipitating it upon their surfaces, which thus become a nucleus around which phosphate concretions are built.

Such phosphate concretions are found on the ocean floor in many localities at moderate depths. They are also found in many old limestones, in which they are generally scattered. By the weathering of this limestone the nodules which are left behind are concentrated into beds which are rich enough to be worked. By solution of the phosphate and by redeposition in veins or on the walls of cavities, or by replacing limestones upon which they rest, rich deposits of phosphate are produced.

Manganese Concretions

Concretions of oxide of manganese with oxide of iron, clay, and other substances are also found on the floor of the deep sea in many localities. The manganese and iron form concentric layers about some nucleus, which may be the tooth of a shark, or some other substance. It is not fully known whether the manganese is derived from the sea-water in which it is present in very small quantities, or whether it is derived from the decomposition of basic volcanic rocks on the floor of the ocean. Manganese concretions are also found in ancient marine deposits, and in some cases at least may represent concentration of scattered nodules by the weathering of the rock which contained them.

Glauconite

Still another deposit formed by chemical means on the sea-floor, is the mineral glauconite, which, when abundant, forms beds chiefly composed of small grains of this mineral. On account of their green colour such beds of glauconite grains are commonly called "green-sands."

Chemically the mineral is an impure hydrous silicate of iron and potassium, and it is commonly formed from fine mud which fills the interior of small shells of Foraminifera, partly by the reaction with the products of decay of the organic matter in these shells, and partly by reaction with the substances in the sea-water. The whole process is too complex, and should be taken up again in a more advanced course.

Beds of green-sands are not uncommon in older marine and other deposits. The most characteristic examples are found in the strata of Cretaceous age which crop out in New Jersey and Maryland. Sometimes by exposure the iron of the green-sand is changed to an oxide, and ochery or red beds will be produced. Such is the vividly red sand bed which is so prominent in the section at Atlantic

Highlands, N. J., and which underlies the town of Red Bank, to which it has given the name. Beds of green-sands are also common in the Cretaceous strata of southern England, where they are generally spoken of as the Green Sands. They also occur in France and elsewhere. Some beds of green-sands are, however, found in deposits which accumulated elsewhere than on the sea-floor.

CHEMICAL DEPOSITS AND EVAPORATION PRODUCTS OF LAKES

LACUSTRINE DEPOSITS

Lakes may be classed as fresh-water, alkaline-water, salt-water, and brine lakes. The salt-water and brine lakes have already been referred to, and it has been shown that the deposits in these are mainly pure salt (sodium chloride), and in some of the larger ones, like Great Salt Lake, also sodium sulphate or mirabilite.

The Caspian must be differentiated from salt lakes of smaller size, as it is more properly a portion of the sea which has been cut off. Hence the depos its there are generally like those formed on the sea-coast.

COMPOSITION OF LAKE WATER

The composition of lake water varies of course in an almost endless manner, no two lakes having water of exactly the same composition. Nevertheless, it is possible to select certain types or averages of groups, which represent in a general way the mineral substances present in such waters. Fresh-water lakes are of course the most abundant, and as their composition varies to a less degree than that of the other lakes, it is possible to give a general average.

This is shown in the first column of the annexed table, in which the average also includes that of river waters, which are not essentially different from those of lakes. In the other columns the composition of a typical alkaline water saline water and brine are given, and the composition of ocean water is again given for comparison. In this table the composition is expressed in terms of ions rather than of salts, which is the more accurate way of statement.

DEPOSITS OF FRESH-WATER LAKES

The three most important types of deposits formed from the waters of fresh-water lakes and ponds are those of carbonate of lime and carbonate and oxide of iron.

Carbonate of Lime

This is by far the most important constituent of lake waters, and for that matter, of fresh-water bodies of all kinds. Nevertheless, the actual quantity is smaller in fresh than in sea-water. As will be seen from the analysis a cubic mile of freshwater contains on the average only 360,915 short tons of calcium carbonate ($CaCO_3$), whereas a cubic mile of sea-water contains 583,520 short tons of this mineral in solution.

It must be remembered, however, that sea-water contains a vastly larger quantity of other substances, of which common salt forms 131,526,000 tons, while in average fresh-water it forms only about 19,656 short tons per cubic mile. Moreover, the total quantity of mineral substances dissolved in the sea is 169,148,000 short tons per cubic mile, in fresh-water only 854,100 short tons; that is, the sea contains about 200 times as much mineral matter in solution.

The mineral matter of freshwater bodies is precipitated by three methods: evaporation, chemical reaction, and organic secretion. The last belongs to the topic of organic deposits.

Evaporation

Lake waters are, as a rule, very far from being saturated with carbonate of lime, and for this reason such a substance will not be deposited as an evaporation product until much or all of the lake water is evaporated. An exception to this is the deposition of carbonate of lime in the form of calcareous tufa in the marginal pools or on beaches of great fresh-water lakes situated in dry climates, but constantly supplied with water by a large river. Such tufa forms by local complete evaporation of the water which lies in shallow marginal pools above the ordinary water-level and is replenished at intervals by spray and the waves.

The spray which saturates the sands of the beaches may also, on rapid evaporation, leave behind carbonate of lime to bind the sand grains together. Such deposits of calcareous tufa are found in the old lake beaches on the slopes of the Colorado or Salton desert up to 40 feet above sea-level. These were formed when the Colorado River drained into the basin and kept it full of water up to that level, in spite of the rapid evaporation which was taking place in the dry climate of the region, and which has completely dried out that basin since the Colorado has become diverted to the Gulf of California.

The nearly complete evaporation of lake water is illustrated by the history of lakes Bonneville and Lahonton, which in an earlier period of the earth's history occupied large areas in the western United States, and the old shorelines of which are still traceable. The product of evaporation of the water of these lakes was chiefly carbonate of lime in the form of calcareous tufa, of which several types were formed.

Significance of Lime Deposits of this Type

Limestones like the above, due to evaporation, indicate a dry climate during the period of their formation. If such limestones are covered by other deposits, which later harden to rocks, they will form a member of a series of stratified rocks similar in general appearance to many that are found in the older parts of the earth's crust.

If in such an older series the limestone member can be determined by its peculiar characteristics to have originated as an evaporation product of an old lake basin, a definite conception of the physical conditions of that region at the time of the formation of the limestone bed is gained. It is therefore important that the particular characteristics of such limestone beds should be understood. At the same time it must be remembered that with the passage of time, such a limestone will undergo more or less change, so that the old characters are to a greater or less degree obliterated or altered.

Nevertheless, enough may remain to indicate the origin of the limestone, and from the adjoining formations and those lying beneath and above it additional evidence for or against the evaporational origin of such a limestone may be obtained. What may prove to be an example of this type formed in a past geological period (Permian), is the bed of Magnesian Limestone exposed on and near the coast of Durham in England, which shows a structure not unlike the spherical structure of the dendritic tufa of Lake Lahonton. The internal structure is, however, more strongly crystalline, which may be due to its greater age.

Chemical Precipitation

This will take place in stagnant lakes and ponds, the waters of which contain much carbonate of lime in solution, and which moreover contain decaying organic matter. This decay, as in the ocean, will produce ammonium carbonate, which will tend toward precipitating the lime.

It may be doubted, however, that much lime is precipitated in this manner from fresh water, since the presence of carbon dioxide, which is also formed as the product of decay, would tend to keep the lime in solution.

Iron Carbonate and Oxides

Stagnant swamps are often covered by an iridescent film, which is the result of the oxidation, on contact with the atmosphere, of iron salts which were contained in the water. Such iron salts are originally in the form of ferrous carbonate ($FeCO_3$), which in mineral form is the ore *siderite*.

This iron carbonate is derived from iron salts in the soil and has resulted from the decomposition of iron-bearing rocks (the ferro-magnesian silicates of igneous rocks), the leaching being performed by rain and ground waters which have taken up carbon dioxide (CO_2) from decaying vegetation. When such a solution of iron carbonate is brought into a lake or swamp, one of two things will happen. When there is much decaying organic matter in the swamp, this will appropriate all the available oxygen, and then the iron is deposited in the form of the carbonate.

This is generally impure, being mixed with the mud held in suspension and carried down with the iron carbonate. This mixture is most frequently deposited around some object which forms the nucleus of an *iron-stone concretion*, as such structures are called. Such iron-stone or siderite concretions when sufficiently pure form ores of iron, and they are commonly found in formations in which coal (the product of the partial decay of the vegetation) is also found.

If, however, vegetation is not abundant, the carbonate will be changed to the oxide, by contact with the air and with the oxygen dissolved in the water, and in such cases an iron oxide will be deposited. This is commonly in the form of limonite, the yellow iron ore, in which water is present, and the loose, porous masses of this mineral which form on the bottom of such ponds are called bog-iron-ore.

Such bog-ores are generally most abundant near the margins of the ponds and swamps and are often wanting near their centers. Sometimes such bog-ore forms very rapidly, some Swedish lakes having deposited layers several inches thick in twenty-six years.

The decomposition of the iron carbonate in the solution and its oxidation to insoluble iron oxide is aided by, and in some cases largely due to, the work of minute organisms in the water, the so-called "iron bacteria." Such deposits might be referred to the organic group, were it not for the difficulty of distinguishing them from purely chemical deposits such as those described.

ALKALINE LAKES AND THEIR DEPOSITS

Under this heading are classed lakes in which carbonate of soda plays a more important part than carbonate of lime, which is generally present in small quantities only. The relatively great preponderance of the CO_2 radical, and the much reduced chlorine content, further differentiate typical alkaline lake waters from those of saline lakes, in which sodium chloride is the dominant salt in solution and the carbonates are insignificant in quantity.

Sulphates and chlorides, however, are generally present, and one or the other or both may be abundant, making complex alkaline waters. In some cases, too, potassium may be an important element, exceeding even the sodium, as in Albert Lake, Oregon.

Different salts may be deposited at different times by these lakes. Thus Searle's Lake or Marsh deposits both carbonate and sulphate of sodium and borax and niter as well. From Owen's Lake, California, both carbonate and sulphate of sodium are obtained. Common salt (sodium chloride) is also an accompaniment of the deposits in many of these lakes.

The greatest deposits of niter known in the world are found in the Atacama and Tarapaca deserts of Chile. The amount has been estimated at 254,760,000 tons and is found at elevations exceeding 2000 feet above the sea and from 50 to 100 miles from the coast. The crude sodium nitrate is known as *caliche* and is associated with anhydrite, gypsum, epsomite, halite, and other minerals.

The origin of these and of the potash nitrates of Bolivia is not fully understood, but they were deposited from solution in drying basins. These lakes deposit chiefly sodium chloride or common salt, especially those in which the waters are a brine, as is the case in most inland salt lakes. Great Salt Lake, Utah, the Dead Sea, and numerous Russian and Siberian lakes serve as examples. They supply vast quantities of common salt for domestic and other purposes.

CHEMICAL DEPOSITS AND EVAPORATION PRODUCTS OF RIVERS

These are of comparatively rare occurrence and are confined chiefly to regions of arid climate. As carbonate of lime is the main mineral constituent of the river water and of that of freshwater lakes, it forms the chief, indeed practically the only, important chemical deposit of fresh water.

In Bahia, Brazil, the rivers which flow in and over the older limestones are highly charged with lime in solution, and under the influence of the tropical sun, partial evaporation of the water and precipitation of the lime takes place. Thus deposits of lime ranging up to 100 feet in thickness have been formed, and they often contain plant remains as well as shells of river and land mollusks of species still living in the region. Angular as well as water-worn fragments of other rock are also included, and at times the mass becomes brecciated through local disruption and cementation of the fragments. Similar deposits are also formed around waterfalls of rivers in limestone regions in various parts of the world.

On the broad flood-plains of many tropical rivers and on their deltas are formed crusts of limestone, which are precipitated from the over-charged river water from the limestone hills. In Mexico and the southern United States these are known as *tepetate,* while the nodular limestone masses embedded with the sediments of the Indus and Ganges, in northern India, are known as *kankar*. Such limestone nodules of chemical origin are very characteristic of river sediments in arid regions, and their occurrence in older sandstones leads us to ascribe a similar origin to these deposits.

DEPOSITS BY SPRINGS AND UNDERGROUND WATERS

Lime Deposits of Springs

In limestone regions, the underground water is generally strongly charged with carbonate of lime in solution. Where this water reaches the surface in springs, the relief of pressure and the escape of carbon dioxide combine to cause the precipitation of some of the carbonate of lime as a flocculent material, which later hardens to solid stone. Where leaves, mosses, or other organisms are bathed by this spring water, they are covered by a deposit of lime or are included as fossils in a mass of calcareous tufa or travertine.

In ordinary spring water, the deposition of lime goes on rather slowly, but sometimes the growth of the travertine deposit is very rapid. At the Baths of San Vignone, in Tuscany, for example, travertine is deposited at the rate of six inches a year, while at San Filippo, in Sicily, the rate is one foot in four months. Here a hill, a mile and a quarter long and a third of a mile broad, has been formed by such deposits, the height being at least 250 feet.

The water of hot springs generally carries more mineral matter in solution than that of cold springs. The cooling of the water on reaching the surface is also very conducive to lime deposition. Hence we have here extensive travertine deposits, the terraces, dams, and basins of the Mammoth Hot Springs of the Yellowstone.

A peculiar form of lime deposit from springs is the well-known onyx marble, or Mexican onyx, which occurs interbedded with normal tufas in Arizona, Mexico, California, and elsewhere in America, and in North Africa, Persia, and elsewhere in the Old World.

This is a compact, highly crystalline, and often beautifully variegated limestone of a semi-translucent character, much used for decorative purposes, the construction of sodawater fountains, etc. It is frequently found resting on crystalline rocks in regions devoid of other limestones, and this suggests that the lime and the water which brought it to the surface were of deep-seated or magmatic origin, that is, derived from hot igneous masses within the crust of the earth.

No beds of this deposit have actually been observed in the process of making, but from the fact that it is generally enclosed by beds of normal tufa, we may assume that these deposits were formed in temporary pools or lakes where standing water prevented the rapid escape of CO_2, the rock thus becoming compact instead of porous, as is the case in ordinary tufa formed on the land.

Oölites and Pisolites Deposited by Springs

At the famous springs of Carlsbad, in Bohemia, carbonate of lime is deposited in the form of spheroidal, discrete masses of the size of a pea, and hence forming an accumulation of particles which when bound together into a rock would constitute a pisolite. As the water rises in the springs it holds in suspension minute mineral fragments such as quartz or feldspar, which then receive a coating of lime precipitated from the water.

As the particles are turned over and over in all directions like a pith-ball in a fountain jet, the coating will be uniform all over, and spheroidal masses are produced. Sometimes gas and air bubbles form the nuclei around which the lime is deposited, thus forming spheroids with a hollow centre. The water of these springs is probably of deep-seated volcanic (magmatic) origin, from which source the lime is also derived. This is indicated by the fact that there are no known beds of limestone through which these waters ascend.

While no doubt both oölites and pisolites have been formed in the past by springs, the great majority of these deposits now found in the rocks of the earth's crust were formed in standing bodies of water through the influence of organic matter either directly, by the physiological activities of living organisms, such as bacteria or minute algæ, or by ammonia generated by decaying organic matter.

LIME DEPOSITS IN CAVES

The best-known types of lime deposits from ground-water are found in caverns. Two types of structures are generally recognized, the stalactite, depending from the roof of the cavern or from some projecting edge, and the stalagmite, which forms on the floor of the cavern, building up a mound or pyramid, or forming a hummocky limestone floor. The formation of stalactites may be observed in tunnels and underground chambers, the roofs of which consist of blocks held together by lime-mortar, as well as in limestone caverns. Percolating ground-water will dissolve a part of this lime, and when a drop of this lime-charged water reaches the tunnel and is suspended from the roof, rapid evaporation will cause the formation, around the drop, of a thin shell of lime.

The pressure of the percolating water will cause the breaking of this film of lime, leaving only a small ring on the roof. The original drop falls, and a new one, equally charged with lime, suspends itself from the bottom of the lime ring. By constant repetition, a slender delicate tube is formed by the successive additions of minute rings of lime, and this is the basis of the stalactite. Other water, running down the outside of this tube, will thicken and strengthen it by the addition of successive layers of lime. In this process the lower end of the initial tube is soon closed, and after that the stalactite remains a solid icicle of carbonate of lime.

Neighboring stalactites may, from close juxtaposition, become confluent and form broad sheets or curtains of lime which are often beautifully banded. Such sheets of lime depending from an edge in the roof or along the line of a crack are shown in Luray Cavern, Virginia, where they form one of the striking features of this beautiful cave.

Stalagmites are built up on the floor of a cavern by the evaporation of the water which drops from the roof and generally from the end of the stalactite. The continued evaporation of this water leaves a minute quantity of lime, which is gradually built up into a mound, and this becomes steeper and steeper as it increases in height, and finally forms a conical or even columnar mass beneath the stalactite, with which it may eventually become joined into a continuous column.

This is well shown in the above view of the interior of Marengo Cave in Indiana. Many such columns may result in the cutting off of chambers and galleries from the original caverns. On the margin of the stalagmite secondary dependent stalactites are often formed, wherever a higher portion of the stalagmite projects cap-like beyond the lower. Thus highly complex and picturesque structures are produced.

Where the water spreads laterally before evaporating, an extended sheet of stalagmite material is formed, and the lateral confluence of many such sheets may result in the production of a stalagmite floor. In many of the limestone caverns of southern France and other parts of Europe, which during Palæolithic time were inhabited by the people of the Old Stone Age, implements and even the bones of these prehistoric people are found embedded in the clay of the cavern floor, over which not infrequently a cover of stalagmitic material has been formed.

Basins are also formed in caverns where water, holding the lime in solution, runs over a ledge. The edges of such a ledge cause the water to break into ripples as it overflows, and this permits the escape of some of the carbon dioxide which holds the lime in solution. In consequence, this lime is deposited at the edges of the ledge, and so a rim is gradually built up, which holds back more and more of the water in a permanent pool. Many such pools are found in limestone caves such as Luray, where the conditions for their formation are favorable. They are analogous to the sinter terraces of the hot springs.

MINERAL VEINS

Of all the deposits formed by waters circulating in or rising through the crust of the earth, the mineral veins are the most important to man. The mineral-bearing waters which form the veins are probably in most cases hot, and they may even be in the form of vapors.

These waters or vapors deposit their load of mineral matter either upon the walls of cavities or fissures, or by replacing the rock material alter it along their passageway, which may be a minute crack or other avenue of escape. The first group is called fissure veins, the second replacement veins or deposits.

Fissure Veins

Deposition in fissures is brought about by the cooling effect produced by the walls upon the solution, or by chemical reaction with the material of the wall rock. Thus an acid solution coming in contact with a limestone would become neutralized and lime salts inclosing other minerals and metallic substances might be deposited. Fissure veins have the form of sheets, or films, of mineral matter cutting the rock. If there is a succession of deposits, the vein will be banded parallel to the wall-rock, and the central portion may be filled with crystals.

Veins of ore minerals or metallic substances contain, besides this material, quantities of other minerals such as quartz, calcite, barite, etc., and these constitute the *gangue material* of the miner. Thus native gold is commonly found in a gangue of quartz, but quartz veins without gold or other metal are much more common. So, too, are veins of calcite and other minerals of little commercial value.

In rocks containing many veins there can often be recognized several distinct series of successive origin. The relative age of veins can be determined from the fact that the younger veins are continuous across the older ones which they intersect. Sometimes cavities of irregular shape in the country rock form the site of mineral and ore deposits, these differing from true fissure veins mainly in their form and extent. Deposits of this kind are called *cavity-filled ore deposits*.

Replacement Deposits

Instead of filling a previously existing fissure or cavity with its deposits, the mineralbearing waters or vapors may deposit their material as they pass through the rock, by dissolving mineral particles of the country rock and filling their places with new mineral matter. Such a replacement of one mineral by another would go on, molecule by molecule, until an area of the rock is so replaced by, or impregnated with, valuable mineral matter as to become an important ore deposit. Frequently, of course, the replacement and fissure type of deposit may intergrade, for the walls of an open fissure may also be partly replaced by mineral matter. Veins in which ores are scattered, *i.e.* lean veins, may become locally enriched by the subsequent solution and local concentration of the valuable mineral matter. The movement of the ores is generally from higher to lower levels. Concentration may also occur by the weathering or solution and removal of the gangue material.

Placer Deposits

Concentration of valuable minerals may also occur at points distant from the original vein. Thus quartz veins carrying gold may be broken up by the weather and the fragments washed away by the streams. On account of the greater weight of the gold, this will be concentrated in favorable areas and so form the well-known *placer deposits*.

Source of the Vein Minerals

The material in solution in the vein-forming waters or vapors may be derived either by solution from the rocks with which surface waters, descending into the earth, come in contact in their circulation through the deeper portions of the earth's crust, deriving their material chiefly from rocks which overlie the point of deposition (descending solution), or the waters and vapors circulating through the rocks laterally may dissolve their ore material and deposit it on reaching the fissure (lateral secretion). Again, water derived as emanation from deep-seated igneous masses (juvenile or magmatic water) may carry upwards in solution the mineral substances derived from these masses and deposit them in the higher fissures (ascending solutions). This last mode of formation is regarded by many as the most typical origin of mineral veins.

Chapter 6

Agents of Erosion

The Loosing and transporting away of rock debris by moving agents operating at the earth's surface is called *erosion*. Weathering of rock is essentially a static process; if there were no natural agents to remove the clays, sands, and other products of weathering, all surface rocks would eventually be deeply buried beneath a thick mantle of their own decomposition products—a common situation on flat ground in areas with a humid climate.

On sloping ground, however, the products of weathering are constantly being transported away by rainwash, rills, and streams. On gentle slopes the soil cover is generally a few feet deep, but on steeper slopes, weathering products are generally removed nearly as rapidly as they are formed. On many steep mountain slopes the erosional agents are so active that they even scour and remove fresh rock before it has been appreciably softened by weathering.

As the products of weathering are removed by erosion, new rock surfaces are exposed to attack by the weathering agents. Thus a cycle of change is set up: Weathering *prepares* the rock for transport by decomposing and disintegrating it. Erosional agencies *transport* the weathered material to a different locality (for example, to the ocean), and there deposit it in layers (strata) which may be ultimately *cemented* into new rock. This rock may then be raised above sea level, exposed to the air and the rains, and again weathered into soil, to start a new cycle.

PROCESS OF EROSION

Erosion is a dynamic process. Only moving agents are capable of picking up rock debris and transporting it to a new location. Although many different agents—as varied in their nature as earthworms, lightning, and avalanches of snow—can transport a minor amount of rock waste, the great bulk of erosional debris is transported by four main agents:

- The *wind*.

- *Waves and currents in the ocean* and, to a lesser extent, in lakes and other small water bodies.
- *Glaciers.*
- *Running water.*

Another important process, because it modifies and increases the erosional effects of running water, is the slow, or occasionally rapid, *downslope movement*, under the action of *gravity*, of solid or semi-liquid masses of mud, soil, and rock debris. On hills and mountainsides such downslope movements deliver a steady increment of rock waste to the streams at the foot of the slopes. The streams then have material to transport.

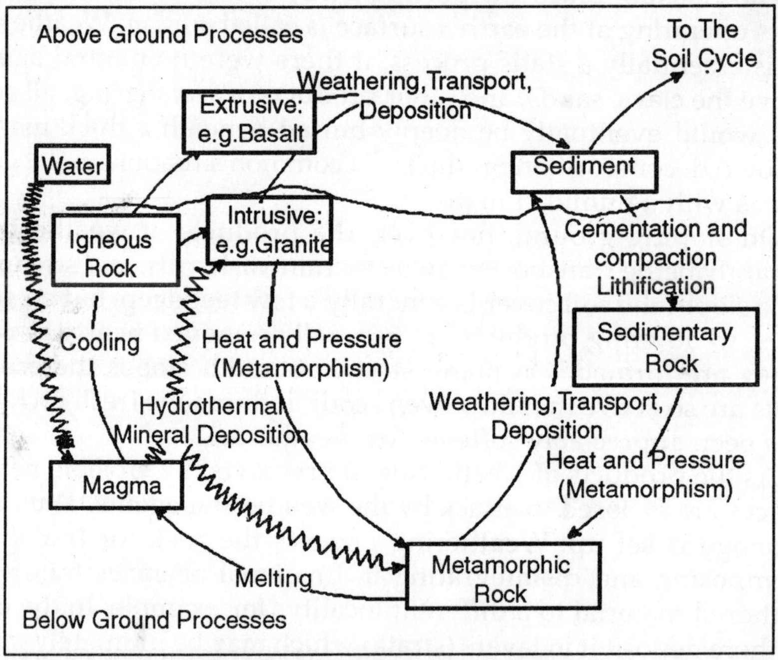

Fig. The Cycle of Rock

To evaluate the effectiveness of the different agents in eroding the land requires analysis of many factors. Because of the size of the earth, the complexity of the erosional processes, and the difficulty of measuring accurately the amount of debris in transit at any one time, it is impossible to get quantitative figures on either the rate at which erosion is lowering the land masses (the *rate of denudation*), or on the amount of debris transported by any one particular agent of erosion.

Yet, despite the incomplete data, fairly reliable estimates as to the relative importance of the different agents of erosion, and a rough but useful approximation of the rate of denudation of the land areas can be made.

WIND

When the Great Plains area east of the Rocky Mountains was settled, the natural grass sod was plowed under, and the light soils were excessively tilled in order to grow corn and wheat. Even before settlement, parts of this area had undergone considerable wind erosion.

The farmers who raced madly into the "Cherokee strip" to stake out homesteads, when the Indian Territory of what is now Oklahoma was opened to settlement, were greeted by blinding dust storms. As more and more land was put under the plow, these "black dusters" became bigger and more frequent. In the early 1930's a succession of dry seasons led to the terrible drought years of 1933 and 1934. Pulverized by tillage and parched by drought, the soil had lost not only its original protective grass cover, but much of its cohesiveness. Thus, the stage was set for one of the most spectacular and destructive events that has occurred in modern agriculture.

On May 12, 1934, strong winds lifted vast quantities of dust from the fields of Kansas, Oklahoma, Texas, Colorado, and other Plains states high into the air and drove it swiftly to the east as a gigantic dust storm. Sweeping swiftly out of the Plains as a blinding, choking mass of particles so dense as to blacken the sky and change day into darkness, the seething dust clouds rolled eastward across the well-watered lands east of the Mississippi River.

Here, where little dust could be gleaned from the forest and grass-covered landscape, the clouds lost some of their density, but they still retained enough solid material to blot out the still as they swirled around the skyscrapers of New York City and out over the Atlantic. Dense, dirty-brown clouds of dust engulfed ships 300 to 500 miles from shore and filtered even into well-protected holds, blanketing the cargoes.

Measurements of the amount of dust in the air, reported by observers from widely scattered points over central and eastern North America, indicate that more than 100 tons of dust per square mile fell in the areas covered by the dust cloud. Since the area affected embraced approximately twothirds of the North American continent and much of the western Atlantic Ocean, it appears that more than 300,000,000 tons of soil was removed from the Great Plains during this storm and strewn over the lands and sea to the east.

Wind strongly winnows the material it transports. Only the finest particles are swirled high in the air and carried far. Most of the material moved by wind is not dust, but consists of coarser grains of silt and sand that roll or skip along the surface of the ground. Studies by soil conservationists indicate that in the Plains country about three-quarters of the soil moved by the wind does not rise into the air but drifts along the surface for a few yards, or at most a few miles, and then accumulates in ditches, around clumps of vegetation, and against fences, buildings, or other obstructions. Thus, for every ton of air-borne dust there are generally two or three tons of coarser debris piled in drifts of sand and silt within or near the source.

The dust storm of May 12, 1934, was followed by many others. How a vast area of the southern Great Plains was converted into a barren "Dust Bowl" with attendant economic disruption and enforced mass migration of people is now an episode of American history.

Because of the great havoc wrought, many of the later dust storms were carefully investigated. One of the most completely studied storms was a much smaller one that swept out of the Dust Bowl on February, 7, 1937. Most of the land to the northeast was blanketed with snow so it was easy to collect and measure the dust that settled. At Ames, Iowa, 34.2 tons of material per square mile was deposited by this storm, 14.9 tons per square mile fell at Marquette, Michigan, and 10 tons per square mile in southern New Hampshire.

More rain and better methods of farming have again brought prosperity to the Dust Bowl, but wind erosion on a minor scale still continues, and grimly threatens a major revival if a series of drought years returns.

Although the great dust storms of the 1930's were largely due to changes in the soil cover induced by man, wind erosion is constantly at work on all surfaces of the continent, and is particularly potent in deserts. In the western Sahara, strong southward-blowing winds (the *Harmattan,* a local name for the Trade Winds) sweep across the desert for about six months of each year.

In exceptionally stormy years they are reported to deposit as much as a foot of silt, dust, and sand along the edge of the desert in northern Nigeria. At times, terrific dust storms also sweep northward across the Mediterranean into Europe. In 1859 such a dust storm covered most of Europe, and deposited 80 to 100 tons of dust per square mile in southern Germany.

OCEANIC WAVES AND CURRENTS

Turbulence in the water envelope of the earth (oceanic waves and currents), like turbulence in the air (winds), is an important agent of erosion. Anyone who has watched the waves of the open ocean crash against the coast need not be told that here is a powerful agent capable of picking up and transporting sand, gravel, and other rock waste. Houses, breakwaters, and seaside roads may be destroyed overnight by the waves, cliffs are undermined, ships driven against the rocks and broken into matchwood, and boulders and sand rolled about like ninepins in the swirling water along the shore.

Sand drifted along the coast during a storm may completely block the mouth of a river, or may be swept far out into deep water. Even on a relatively quiet day, the sand along an ocean beach is in constant agitation: each oncoming wave, as it strikes the beach, brings sand rolling along the bottom with it, and as the wave ebbs the sand is rolled back seaward. During severe storms huge boulders may be dashed back and forth in the breakers, and even hurled against a sloping rock beach with such violence as to catapult high in the air with the spray. Occasionally, shore-driven rocks are thrown over seawalls and come crashing through the roofs of buildings.

Obviously the shore of the open ocean is a zone of vigorous erosion. Most ocean coasts are cliffed because the waves rapidly undermine the shore. Along many shorelines, such striking changes take place during storms that valuable beach property ordinarily must be protected from erosion (or in some places from unwanted deposits of new sand or other beach materials) by breakwaters.

GLACIERS

Probably the Nisqually glacier, which is easily reached by highway in Rainier National Park, has been visited by more people than any other in North America. The Nisqually is a pygmy among glaciers; less than 5 miles long and less than 1,000 feet thick, it does not compare with the much larger Emmons and Winthrop glaciers in the more remote northeastern section of Rainier National Park, not to mention the 60-mile-long ice streams of Alaska or the vast moving ice sheets that blanket Greenland and Antarctica.

Dwarf though it is among its fellows, no visitor can examine the front of the Nisqually glacier without being impressed by the tremendous erosive power of moving ice masses. At the glacier front very little ice is visible-most of it is obscured by blocks of rock riding on the moving ice mass or pushed forward in front of it. Some of these blocks are fragments of andesite, each weighing several tons.

By matching the rock with outcrops far up the valley, it is obvious that they have been torn from parent rock ledges upstream and carried to their present resting place. Some have been rafted along on the back of the glacier, some were frozen into its body and dragged forward with it, and some were pushed in front of it. Smaller particles of rock and soil are mixed with the larger boulders.

As the ice melts, they accumulate in hummocky piles in front of the glacier. Thick accumulations of loose rock also rest on top of the glacier, especially at its edges, because, here, boulders, loosened by frost action from the cliffs above, are constantly rolling down, and snowslides and streams of meltwater also add debris to the glacier's surface.

The Nisqually glacier has been slowly shrinking in recent years. Downstream, below the front of the glacier, lie irregular piles of debris left by the ice as it retreated upstream. The rock floor, where bared by the melting ice, is scored and polished by the blocks of rock and finer material that has been dragged across it by the moving glacier.

Meltwater pouring from tunnels in the Nisqually glacier is a dirty milky white. Examination shows that it is cloudy, not from clay, but from pulverized rock flour obviously made by the crushing and milling of rock particles against one another and against the bedrock floor as they were rolled beneath the tremendous weight of the moving glacier and dragged forward by it.

Relative Importance of Different Agents

Spectacular as the results of wind, ocean, and glacier appear, they are of relatively little importance as compared with the movement of rock debris by running water.

In the years since the middle 1930's, rainwash has almost completely obliterated the drifts of silt and sand made during the dust storm of May 12, 1934. Only in the driest deserts are features due to wind erosion conspicuous, and even here the presence of numerous gullied slopes, dry stream beds, and sheets of water-deposited detritus shows that the rare rainstorms are able to put their stamp upon the desert landscape.

Oceans are limited in scope and effect. The attack of the sea upon the land is chiefly confined to the shoreline and to the shallow submerged margin of the continents. Here they have undoubtedly played a part in extending the upper of the two dominant surface levels, but stream deposits have doubtless played an even greater role.

Again, compared to running water, glaciers are quantitatively unimportant as erosional agents. Streams are practically universal—there is scarcely any part of the continents that is not subject to at least some rainwash each year. Glaciers, on the other hand, are confined to high mountain ranges and to polar regions. The combined area of all the glaciers constitutes only ten per cent of the land area. Their contribution to the denudation of the continents is insignificant when compared to the work of streams.

Streams, aided by downslope movements, are the great levelers. Although individual rivers during stage of flood may produce sudden and spectacular feats of erosion, the quantitative importance of streams is due to the fact that rain falls on all parts of the continental masses, and in coursing off them to the sea it steadily removes rock debris from the surface. Let us turn our attention to an analysis of the way in which streams perform their work.

ANALYSIS OF STREAM EROSION

Source of the Runoff

"From whence springs the water of the rivers?" is a question to which reliable answers have been won only in comparatively modern times. The early philosophers held that rainfall was totally inadequate to account for the vast flow of water in rivers like the Seine, Rhône, and Nile, and that the earth's surface was too impervious to allow the percolation of rain water into the soil and rocks from which it could be returned in the form of springs. Although there were occasional dissensions, the general belief up to the middle of the. Seventeenth Century was that the water rising in springs and flowing to the ocean in streams could not possibly be the runoff from rain and snow.

The modern concept of hydrology (the science that treats of water) began with the work of a Frenchman, Pierre Perrault (1608-1680). Perrault made measurements of the rainfall in a part of the Seine River basin over a threeyear interval and also measured the amount of water flowing in the Seine. From the available maps, he then estimated the area of the river drainage basin above the point where he had measured its flow.

These measurements indicated that the Seine carried off only one-sixth of the water that fell within its drainage basin as rain and snow. Perrault thus proved the fallacy of the old assumption that the streams carry more water to the sea than falls on their drainage basins. In fact, the problem of the streams was now reversed—the question was no longer, What is the source of the water? but, What has become of the vast volume of water, equivalent to five times the flow of the Seine, that has been precipitated on the drainage basin but is not being carried out by stream flow?

A few years after Perrault's measurements, Edmund Halley, an English astronomer, showed by experiment that the moisture evaporated from the surface of the oceans is entirely adequate to supply the water returned as runoff.

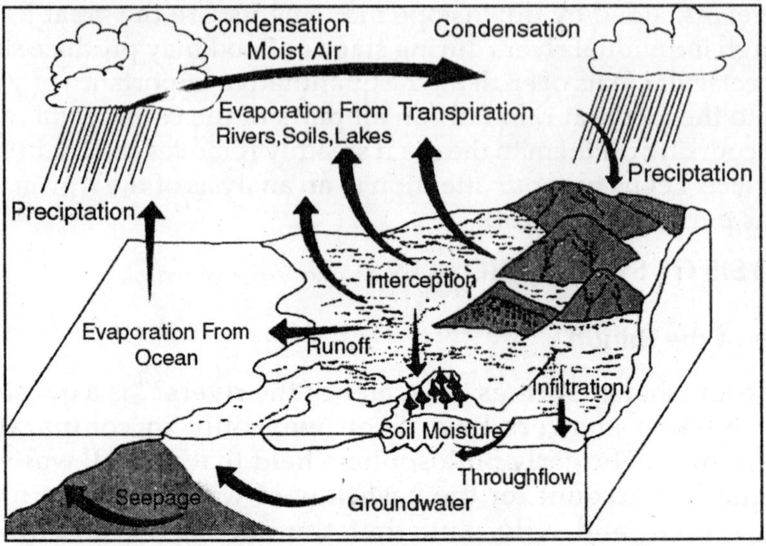

Fig. The Hydrologic Cycle.

Since the time of these pioneers a wealth of hydrologic observations has become available. The flow of water in the major streams of practically all countries has been accurately determined by stream gaging. Such measurement, generally carried on by federal or state agencies, has many purposes. Precise data on river flow are valuable to navigation, irrigation, hydroelectric power, flood control, and many other interests.

Similarly, the accurate recording of rainfall and snowpack has many uses; and data are available from thousands of stations scattered all over the world. Numerous tests have been made on the rate of evaporation from bodies of water under different climatic conditions, and on the rate of loss of water into the atmosphere through transpiration by plants.

Rates of underground percolation of water and measurements of the amount of water stored in underground reservoirs have been obtained from thousands of observations on springs and wells. From these measurements and observations we can fairly accurately deduce the relations between rainfall and runoff, and can answer the question of what happens to the water that falls as rain.

The Hydrologic Cycle

Of the moisture precipitated on the land as rain and snow:

- Some evaporates at once from the surface of the ground and from the vegetation on which it falls;
- Some is absorbed by the roots of growing plants and is quickly transpired back into the atmosphere through their leaves, although a little enters into the plant tissues and is trapped there until the plant is dead and destroyed:
- Some seeps into the soil and rock where it may be temporarily stored underground until it is:
  - Brought to or near the surface by capillarity and evaporates;
  - Delivered to the roots of growing plants;
  - Oozed to the surface at a lower elevation, reappearing either in springs or as ground-water runoff (*effluent seepage*) pouring into streams along their banks and bottoms;
- Some runs off in surface rills, brooks, and rivers and is returned directly to the ocean. It is the last part, the *runoff,* that we are concerned with in the study of stream erosion.

The hydrologic cycle is depicted graphically. The sun is the source of energy that operates this great system of waterworks. Solar energy raises the temperature of the oceans and the lands, and evaporates water from their surfaces. It also stimulates the growth of plants and leads to the transpiration of water vapour into the atmosphere from their pores. Atmospheric moisture gained in these ways is wafted inland and raised to high altitudes by the winds.

It thus acquires potential energy of position. When the moisture falls as rain, this energy of position is converted to kinetic energy—the water has become a moving agent, capable of acquiring and moving a load. On reaching the surface of the earth, that part of the rain that constitutes the runoff flows downslope in response to gravity, moving obstacles in its path and sweeping soil and rock debris along in its currents.

In studies of the relation of runoff to precipitation it is convenient to define *runoff as the total discharge of water by surface streams.* Thus defined, the runoff includes not only that part of the precipitation that flows directly from the rains across the surface of the land, but also the increments of ground water that enter the streams from springs and effluent seepage.

It is also convenient to distinguish that part of the precipitation returned to the air by evaporation and transpiration as the *evapotranspiration factor*. Hydrologists, for convenience, usually simplify the relations within the hydrologic cycle by assuming that evapotranspiration can be determined by subtracting runoff from precipitation. The equation

Precipitation = Runoff + Evapo-transpiration

is only approximately correct, however. It neglects such locally important (though generally unmeasurable) factors as the seepage of water through the soils and rocks directly into the ocean. For the land masses as a whole such seepage is not great, but in a few areas of highly permeable rocks, as in the Hawaiian Islands, it may locally dispose of a large part of the total rainfall. Other very minor factors are the amounts of water locked up in the tissues of plants and animals, and the amounts chemically combined in minerals during weathering.

VARIABLE FACTORS AFFECTING THE RUNOFF

The ratio of runoff to precipitation is not everywhere 1:6 as Perrault determined for the basin of the Seine. Several variable factors affect the ratio.

Careful studies of many different hydrologic basins in the United States, made by the Water Resources Branch of the U. S. Geological Survey, have revealed great variations in the ratio of runoff to precipitation. From these measurements some of the factors may be summarized as follows:

Temperature

Evaporation is, of course, much greater in warm regions than in cold. High temperatures also favour growth of vegetation, and hence increased loss by transpiration. In humid regions a close correlation has been found between the amount of runoff and the mean annual temperature. Even in arid and semi-arid regions, if large groups of data are considered, not simply those from a single drainage basin, there is also a reasonably close accord.

The data relating temperature to runoff in the United States has been summarized by Langbein, a hydrologist of the U. S. Geological Survey. Thus, for an average annual precipitation of 40 inches, more than half the rainfall (21.5 inches) runs off where the mean annual temperature is 40° F.; but the runoff decreases to 10.2 inches in a mean annual temperature of 60° F., and to less than 3 inches in a mean annual temperature of 80° F.

Slope

Slope obviously exerts an important control on runoff. Steep slopes of bare rock in mountainous regions shed practically all of the rain that falls on them. On flat ground the rain may remain on the surface in shallow puddles until it is evaporated or absorbed by the soil and plants.

Permeability of the Ground

Different soils and rocks have highly different permeabilities. Most of the rain falling on the porous ash of a recent volcanic cone sinks directly into the ground and becomes a part of the ground-water circulation, but nearly all will run off on a landmass of similar slope that is underlain by relatively impermeable shale.

The ability of a soil to absorb rainfall or snowmelt depends on other physical factors as well as permeability. Water-logged soil, although its permeability may be high, has reached its maximum capacity of absorption; frozen ground is similarly nonabsorptive. The absorption of a highly permeable soil also may be limited because of a nonpermeable subsoil or bedrock a few feet below. Permeability, then, is not alone an accurate measure of the rate of infiltration.

Hydrologists use the term *infiltration capacity* to define *the maximum rate at which the soil, in a given physical condition, can absorb falling rain.* Artificial tests of infiltration capacity made by playing sprinklers on enclosed plots of soil from which the runoff is caught and measured show, as would be expected, that the rate of absorption is high at the beginning of a rain, then diminishes rapidly until a fairly constant value is reached. The high initial rate usually changes to the slower uniform rate within a half hour after the rain begins to fall.

Amount and Duration of the Rainfall

Distribution of rainfall through the year greatly influences the runoff. Rain uniformly distributed in many small showers may be largely evaporated or absorbed by the ground and plants before it reaches flowing streams. During violent rainstorms, on the other hand, percolation is too slow to lead much of the water underground; most of it courses rapidly across the surface and into stream channels. Rapid melting of the winter snowpack by warm rains or winds is one of the most common causes of river flooding.

If the annual rainfall is high (80 inches or more), most of it is likely to run off, even though the rain may be uniformly distributed. Here, the underground reservoirs are constantly replenished, and the pores within the soil and rock are filled with water almost to the surface of the ground. In deserts, on the other hand, most of the scanty rainfall is absorbed by the parched soil, or evaporated, and there is little runoff except immediately after violent storms.

Vegetation

Vegetation retards the runoff. Tangled stalks of grass or the mulch of decaying leaves and twigs in a forest absorb rain like a blotter. Earthworms and other burrowing animals that live in the plant-rich soil aid percolation by opening tunnels in the soil. Very striking increases in runoff have been observed where the forest cover has been burned off, or where the natural sod has been plowed under, as in our Dust Bowl. summarizes some of the data for a few such areas.

These variable factors result in marked differences in runoff in different stream basins. In the western United States runoff varies tremendously with differences in temperature, infiltration capacity, and other local factors. The range is from less than 0.25 inch in the southwestern Arizona desert to more than 80 inches on the western slopes of the Olympic and Cascade mountains of Washington and Oregon.

AMOUNT OF WATER AVAILABLE FOR EROSION

The rivers of the United States, as determined by stream gaging, deliver water to the oceans at a rate of 1,800,000 cubic feet per second (about 330 cubic miles per year). Roughly one-third of this amount is carried by the Mississippi River.

The average annual rainfall of the United States, determined from thousands of rain-gaging stations, is 30 inches. The average runoff for the country as a whole is 8.6 inches. This gives a ratio of runoff to precipitation of approximately 1 to 3.5, as compared with the ratio of 1 to 6 determined by Perrault for the basin of the upper Seine.

Data for runoff and precipitation of other continents is less complete than for the United States. L'vovich, a Russian hydrologist, has summarized the available data on runoff for the various continents. The table shows that Australia, with a runoff of only 3.0 inches, is the driest of the continents, and that South America, with 17.7 inches of runoff, is the wettest.

The world average, as given by L'vovich, is 10.5. Other hydrologists have reached slightly different, and in general somewhat lower, for the average world runoff. Despite these uncertainties, we can make a fairly reliable estimate of the total amount of water available for erosion: The average annual precipitation over the land areas of the earth is about 40 inches per year.

This means that each year approximately 35,000 cubic miles of water falls on the land in the form of rain and snow. According to different estimates, the rivers return to the oceans from 7.5 to 10.5 inches per year. Using an average figure, this means that in round numbers approximately 8,000 cubic miles of water courses off the lands and into the seas each year. This is the amount of water that is available for the erosion of the lands.

ENERGY RELEASED BY THE RUNOFF

On the average, the lands stand about one-half mile above sea level. Therefore, the continental runoff descends from an average altitude of one-half mile to sea level in flowing from its source to the ocean. This is an average figure; runoff falling on a plain near sea level may descend only a few feet to reach the ocean; the meltwater from the snows on the summit of Mount Everest falls over five miles.

A tremendous amount of energy is developed by the world's streams during their descent to the sea. Some qualitative idea of the amount can be gained by imagining all the continental runoff to be concentrated in one gigantic river that is pouring over a huge waterfall equivalent in height to the average altitude of the landmasses.

In other words, imagine 8,000 cubic miles of water per year cascading down a waterfall one-half mile high! One could easily compute the amount of power available at such a waterfall, but the resulting figure is so large in human terms as to be utterly meaningless.

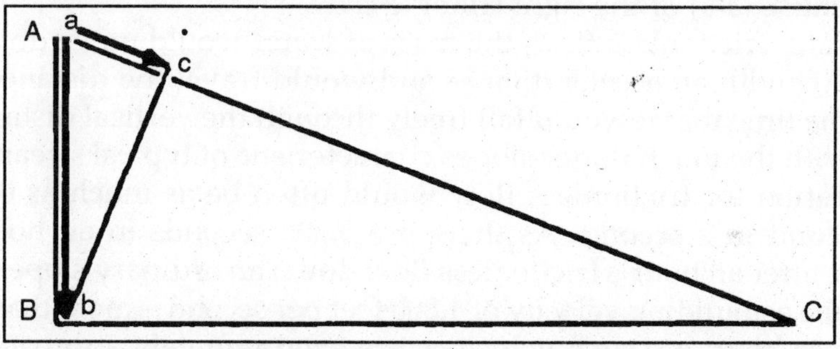

Fig. Resolution of Gravity for a Stream of Uniform Slope.

Yet this is the amount of energy available for the erosion of the lands by streams. It is easy to see why running water is the great leveler—much more important than all the other agents of erosion combined. Fortunately for mankind, very few streams work at anywhere near their full erosional capacity.

Because weathered material ready for transport is moved downslope to the streams only slowly, many rivers carry little load and dissipate most of their energy in work other than erosion. Otherwise, the continents would be largely denuded of soil.

MECHANICS OF STREAM EROSION

In dealing with influences on stream flow it is essential to keep in mind the *Law of Conservation of Energy*, and the distinctions between potential energy (energy of position), kinetic energy (energy of motion), and heat energy. The *Law of Conservation of Energy* requires that, although one form of energy may be transformed into another, for example, kinetic energy into heat energy when a falling rock strikes the ground, the sum of all the different kinds of energy does not change during the transformation. In other words, the total amount of energy remains constant, and no energy is gained or lost.

Flowing streams are powered by gravity. It is convenient to use a simple graphical representation, the "resolution of forces", to determine the component of the force of gravity effective down the particular slope on which a stream is flowing. Consider a mass of water poised at the top of a uniform slope. Its total energy is measured by the potential energy of vertical fall, and hence the final velocity after movement down the slope would, if there were no friction, be the same as for direct vertical fall, though the time required to attain the velocity would be greater because the component of gravity operating parallel with the slope is less. The acceleration of gravity (an increase in velocity of 32 feet per second at the end of each second) is represented by the hypotenuse (*a-b*) of the right triangle *abc*.

Under frictionless flow, the mass of water would move down the slope AC with an acceleration *ac* and would travel the distance *ac* in the same time that it would fall freely through the vertical distance *ab*. Even with the much flatter slopes characteristic of typical streams, the acceleration for frictionless flow would often be as much as 0.5 feet per second in a second. As there are 3,600 seconds in an hour, the stream, after an hour's frictionless flow down an ordinary slope, might attain the astounding velocity of 1,800 feet per second—more than 1,200 miles per hour! No such velocities are even remotely approached in natural streams. Actual streams are slowed by friction of the moving particles against one another and against the land surface.

They commonly attain and maintain velocities of only a few feet per second (generally less than 5 miles per hour); if their velocity changes appreciably, it usually decreases downstream. We conclude that the energy of fall is mostly converted to heat energy by friction. Some of the energy, is also used tip in the transportation of eroded rock particles from upstream. The actual quantity of mud, sand, and gravel eroded varies widely for different rivers. For example, compare the crystal-clear Columbia River at Wenatchee, Washington, with the Missouri River (often called the "Big Muddy") near Saint Louis, Missouri. The amount of water carried by each is about the same, but the Columbia, though a roily turbulent stream filled with violent rollers and eddies, contains very little silt and mud. It is flowing on smooth bedrock and coarse boulders, hence almost no fine debris is available in its bed to be picked up and carried. By contrast, the Missouri is laden with mud and silt. It flows across the Great Plains, where the soil cover is deep and the vegetation scanty; consequently, abundant fine-grained rock waste is available for transport.

Much of the energy of flow of the Columbia is used up in internal friction within its boils and eddies, whereas in the Missouri much of the available energy is used to transport the abundant rock waste. But even the muddy Missouri is transporting far less material than it might. Only rarely does the Missouri carry more than 20,000 parts per million (2%) by weight of solid material. This is a high load for a big river, though it is exceeded by the Colorado, the Yellow River of China, and several others. The Mississippi, generally considered a very muddy stream, carries only about 5,000 parts per million (0.5%). Small streams in semi-arid regions with a thick soil mantle, however, may carry enormously greater loads—up to 30 per cent or more by weight—especially during floods. A sample from the San Juan River in Colorado when it was in flood contained over 75 per cent by weight of red silt and sand. There is, of course, a complete gradation from such highly loaded streams to mudflows and landslides. The evidence indicates, however, that nearly all streams are working far below their potential capacity as erosional agents. If they used most of their energy in carrying rock waste, few slopes could retain a cover of soil.

RATE OF DENUDATION

What, then, are the actual figures on the rate of denudation? At the present time, measurements on the rate at which different drainage basins are being lowered are so few and incomplete, and quantitative data so hard to obtain, that estimates can be only tentative.

The total load of debris being carried by a stream is difficult to measure. Although reliable figures covering a period of more than a year or two are available for only very few rivers, there are enough data on hand to indicate the general magnitude of erosion. A carefully controlled series of measurements of the silt load of the Missouri River above Kansas City made during 1930 and 1931 indicates that this river is lowering its drainage basin about 1 inch in 650 years, or 1 foot in 7,800 years.

The Colorado River, which, in the ten-year period preceding September 30, 1935, carried an average annual load of 250,000,000 tons of silt and sand past the Grand Canyon gaging station, is certainly lowering its drainage basin more rapidly than the Missouri; but the

Columbia, at least above the junction with its muddy tributary, the Snake, is undoubtedly denuding its basin much more slowly than the Missouri. Each year the Mississippi River brings to the Gulf of Mexico about 730,000,000 tons of dissolved and solid material.

From such measurements it is estimated that *the rate of denudation for the entire United States is about 1 foot in 9,000 years*. If this rate could be maintained, and if there were no compensating upward movements of parts of the earth's crust, all the landmasses would be eroded to sea level in about 23 million years.

Changes in the Rate of Erosion

Because of certain variables not taken into account in the measurements, our figure of 1 foot in 9,000 years for the average rate of denudation must be modified considerably in dealing with specific erosional problems. For some kinds of problems the figure is too high; for others it is much too low. For example, can this figure be used as the average rate of erosion for all of geologic time? There are many reasons for believing that present-day streams carry much higher loads than the average for the geologic past. The activities of civilized man contribute vastly to present stream loads. Rivers in populous areas carry a vastly increased load because they are polluted with industrial and municipal wastes. Enormously increased amounts of silt and mud are also being washed into our streams from cultivated fields; before human settlement, the original forest and grass cover surely impeded erosion and greatly lessened the amount of soil washed to the sea.

Such an increased rate of erosion has been proven recently by the Soil Conservation Service of the U. S. Department of Agriculture in making extensive tests of the amount of soil removed under varying conditions of tillage. The bureau caught and measured carefully the amount of soil and water running offfrom plots of ground with identical areas and slopes but with different vegetative covers.

These factors cannot be precisely evaluated and, thus, the rate of erosion during the whole of geologic time cannot be fixed quantitatively. Certainly, the rate of 1 foot in 9,000 years is much too high—perhaps two or three times too high. An ingenious attempt to measure the age of the oceans by their content of salt failed because the amount of dissolved salt carried to the sea by the rivers of the geologic past could not be accurately estimated—the age, as determined from the present rate, was many times too low. Can we apply the rate of 1 foot per 9,000 years to the problem of soil erosion? The soil conservationist wants to know how rapidly the topsoils of American farms are being gullied and rendered unusable by erosion.

In many sections of the country productive topsoil is rapidly disappearing, and once-prosperous farms are now barren wastes of deeply eroded gullies and exposed bedrock.

It is estimated that about 282 million acres of farmland in the United States have been so seriously eroded as to make farming unprofitable. Still another 775 million acres have lost enough topsoil to affect their productivity; and another billion acres of land show some evidence of damaging erosion. These figures do not include the large acreage of the Dust Bowl, in which the land has been damaged by wind erosion. The soil conservationist has difficulty in reconciling this striking evidence of damaging erosion with the figure of 1 foot in 9,000 years for the rate of denudation. Most of our productive soils are at least 1 foot thick; how can so much damage have been accomplished in the relatively short time since the sod of the frontier was plowed into farmland? Only after intensive investigations of the kind represented thousands of soil plots scattered all over the United States did the answer become clear. Most of the soil removed from the fields by rainwash is almost immediately redeposited; only a part is transported directly to the sea. In the words of Hugh Bennett, an American soil scientist with the Department of Agriculture:

The sediment entering the oceans represents merely a fraction of the soil washed out of the fields and pastures. The greater part is piled up or temporarily lodged along lower slopes, often damaging the soil beneath; or it is deposited over rich, alluvial stream bottoms or in channelways, harbors, reservoirs, irrigation ditches, and drainage canals... Available measurements indicate that at least 3,000,000,000 tons of solid material is washed out of the fields and pastures of America every year. Bennett's figure for the tonnage lost from the fields is nearly three times the amount of soil delivered to the oceans by streams. Evidently, the figure of 1 foot per 9,000 years for the rate of denudation, though a reasonable estimate of total lowering of the continent, does not take into account the *local* erosion and redeposition of soils of primary interest to the soil conservationist.

GEOLOGIC EVIDENCE OF EROSION

Most of the data on which is based have been accumulated during the last one hundred years, and much of it within the last fifteen or twenty. Since the earliest days of the science, however, geologists have been aware of the vast changes in the landscape brought about by running water. Their observations, that indicated an immense amount of erosion of the land surfaces, were not concerned with the amount of silt being carried to the sea in streams, nor even the spectacular effects occasionally produced by catastrophic river floods, but with the etched surface of the earth itself. On the walls of the Grand Canyon of the Colorado River appear horizontal layers of limestone, shale, and sandstone. The limestones, particularly the thick Redwall limestone which occurs 1,000 feet or more below the canyon rim, form great cliffs. These cliffs can be traced in and out of the short tributary canyons that score the walls of the Grand Canyon.

The cliffed limestone lies everywhere at the same level, and is exposed on both walls of the canyon and its tributaries, and in all the isolated buttes within the canyon that rise high enough to intersect it. Clearly, the limestone must once have been continuous across the canyon; the Colorado River has cut down through the horizontal layers, and eroded away the portions of them that formerly filled the area now occupied by the canyon. At the bottom of the Grand Canyon the river is cutting into granite and metamorphic rocks. From their mode of origin, these rocks must have been formed far below the surface of the ground. In many mountain ranges, strata of sandstone, shale, and limestone have been tilted from their original horizontal position. The edges of these tilted beds, laid bare by erosion, can be followed in ridges. Obviously, the deeper and older beds in the sequence could only have been exposed to view through the removal by erosion of the younger overlying beds that once covered them. In several mountain chains such series of tilted beds are exposed in belts many miles wide; to strip away all of the overlying beds and to expose the deepest layers in some of these ranges have required the removal of 5, 10, or even more miles of overlying rock.The study of this phase of erosion, however, can best be deferred until we have learned about geologic maps, for the most striking proofs of deep erosion have come from the mapping of the surface geology of mountain ranges.

OTHER IMPLICATIONS OF EROSION

There are other interesting implications that follow from our study of erosion. Streams, glaciers, and waves all move material downhill; it ultimately comes to rest at or slightly below sea level. Can this be one factor in accounting for the upper of the two dominant levels of the earth's surface ?What isostatic response can we expect in the earth's crust as a result of the unloading of mountainous areas by erosion, and the loading of the continental shelves and other low areas by deposition? Answers to these questions must be deferred, but the point emphasized here is that erosion is a process that is continuously operating to upset the condition of isostatic balance in the earth.

Chapter 7

Nature of Stream

EROSION AND DEPOSITION

The sediment load of rivers and smaller streams provides undeniable evidence of the erosive power of running water. Surface runoff picks up and transports small amounts of mud and sand from its first confluence in rills and tiny brooks, and eventually enlarges these miniature valleys into gullies.

Rills join to form small streams, and these, in turn, unite into rivers. The whole system erodes the surface of the land and transports the debris to the ocean. Knowledge of how streams erode, transport, and deposit rock waste is essential to an understanding of the development of our present landscapes and to the interpretation of the origin of ancient sedimentary deposits.

Furthermore, the study of stream flow contributes to the efficient use of rivers by man. Without a knowledge of stream mechanics we cannot hope to solve the problems of recurring flood disasters, erosion of rich valley soils, water storage, and river navigation.

River Systems

Geologists and engineers often use the term *stream* in a general sense, that is, to designate bodies of running water of any size. For instance, the term *stream erosion* refers to the action of rivulets as well as rivers.

A *river system* includes the various streams that join within a drainage area. A river or *trunk stream* is formed by the confluence of *tributary streams,* themselves formed by smaller tributaries, and so on to the individual contributions of springs, rills, and rainwash.

Several characteristics of river systems change fairly regularly from the smallest headwater tributary to the mouth of the main trunk river. The most obvious variables are:

- The *gradient* of the stream, commonly measured by *the number of feet of fall per mile of course.*

- The form of the stream valley.
- The amount of erosion or deposition along the course.

Most upper tributary streams flow with steep gradients (commonly over 50 feet per mile) through topography marked by valleys and ridges. They actively erode the bedrock along their courses. On the other hand, the gradient of a trunk stream, especially in its lower part, is ordinarily less than 2 feet per mile. Here deposition equals or may even exceed erosion and the river flows in a sinuous course upon a smooth plain of its own deposits.

It is along the lower, gently sloping parts of river systems that the problems of stream control are most keenly felt. It is here, also, that man uses rivers for navigation, builds great cities, and tills the richest of farmlands.

The relations of rivers to the valleys in which they flow have long been recognized. In 1802 the Scotch mathematician and naturalist John Playfair observed the nice adjustments within a river system and concluded that rivers must surely have cut their valleys:

Every river appears to consist of a main trunk, fed from a variety of branches, each running in a valley proportional to its size, and all of them together forming a system of vallies, communicating with one another, and having such a nice adjustment of their declivities, that none of them joins the principal valley, either on too high or too low a level, a circumstance that would be infinitely improbable, if each of these vallies were not the work of the stream that flows in it.

Today, with a thousandfold more data than were available to Playfair, it is found that stream flow, erosion, and deposition are in extremely sensitive balance. In evaluating this balance quantitatively, geologists and hydraulic engineers have recognized several variable factors. Among these are the gradient and shape of the channel; the velocity of the stream, including variations of velocity within the channel; and the nature of the material that is being eroded and transported by the stream. Before these variables even qualitatively, however, it is necessary to understand the nature of fluid motion itself.

LAMINAR FLOW AND TURBULENT FLOW

If needle-thin streams of colored fluid are admitted to a stream of water flowing very slowly through a glass tube, the colored threads advance in parallel lines downstream from the points of contamination. Where an obstruction is met, *left*, the flow lines move around the object in a "streamlined" way.

This phenomenon, called *laminar flow,* is apparently produced by the sliding of thin layers of fluid over one another. It is characteristic of viscous liquids, such as cold lubricating oils or sticky molten lava. It can be produced in water only at velocities less than an inch or so per second (1/16 mile per hour).

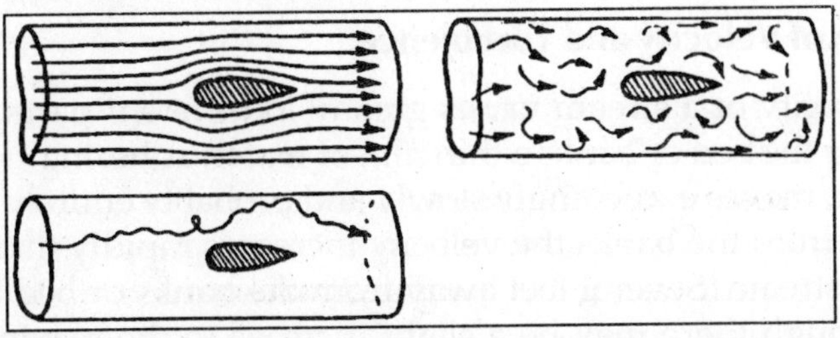

Fig. Left, Laminar Flow; Centre and right, Turbulent Flow.

If the flow is speeded up, the pattern of colored threads abruptly changes. Instead of parallel lines they begin to trace irregular, tortuous paths, then swirl in all directions to the extent of losing their identity by mixing in the water stream. This is called *turbulent flow.* As the velocity of the stream increases, or if the walls of the tube are roughened, the turbulence is increased.

VELOCITIES AND TURBULENCE OF NATURAL STREAMS

Many measurements show that nearly all natural streams, even the smoothest flowing, move rapidly enough to produce turbulent flow. The power of a stream to erode and transport apparently depends on this turbulence, and also on the distribution of velocities and turbulence within its channel.

Range of Velocities

Stream velocities range widely, depending mainly on the *gradient,* the *shape of the channel,* and the *discharge,* that is, the quantity of water passing a given point, usually measured in cubic feet per second. The effect of channel shape on velocity is well shown by the Rhine, which flows at a rate of 2 to 3 feet per second in the broad, flat channel between Basel and Bingen, whereas in the narrow, deep gorge guarded by the Lorelei, the velocity increases to 6 to 10 feet per second.

The effect of discharge on velocity is strikingly evident during floods. Mississippi River velocities measured near Vicksburg in 1937 and 1938 were mostly between 2 and 7 feet per second, but increased to 10 and 14 feet per second during the February, 1937, flood.

The Columbia River, about 170 miles from its mouth, had a mean velocity of 1.9 feet per second on March 12, 1945, when the discharge was only 78,000 cubic feet per second, but a mean velocity of 11 feet per second at flood flow of 1,018,000 cubic feet per second on May 3,

1948. Where stream gradients are very steep, as in mountain torrents, velocities in excess of 30 feet per second (more than 20 miles per hour) have been measured.

Distribution of Velocity and Turbulence

The velocity of a stream varies greatly in different parts of its cross section. Near the bed or banks a thin film of water, adhering closely to the solid surface, moves exceedingly slowly and probably entirely by laminar flow. Away from the banks the velocity increases rapidly through a foot or so of the stream. Several feet away from the banks or bed the current is nearly though there may be a slight increase to the maximum in the stream centre. Turbulent flow is most marked in the zone of sharp velocity change, bordering the stream's channel. Spiraling and rolling eddy currents are thrown from this layer of unbalanced flow into the main body of the stream. The regions of most prominent turbulence are shown in their relation to the velocity distribution.

HOW A STREAM CARRIES ITS LOAD

The Suspended Load

The turbulence in streams accounts largely for their ability to carry a load of silt and clay particles. Whether the particles are brought by rainwash and rills or are eroded directly from the stream's bed or banks, it is the *upward* currents of turbulence that allow their transport. Once a grain has been lifted by an upsweeping eddy, it is kept from falling out again only by the force of other upward moving filaments. True, there are as many downward currents as upward, but both are randomly distributed, so a particle may long be sustained before striking a downward swirl.

Even if the grain is eventually dropped, or carried down in an eddy, other grains will be picked up by currents acting in the opposite direction. The important result is that, in the time interval during which the grain is picked up, jostled around, and deposited again, it has moved downstream a distance that depends on the average forward velocity of the water. To an underwater observer moving with the current, the grains will appear to rise, gyrate, bob, and fall; but to an observer on the bank the stream appears a homogeneous mixture of water and sediment, all flowing downstream. Because turbulent currents seem actually to suspend sediment in the stream, this part of the stream's load is called the *suspended* load.

The Bed Load

Even-flowing rivers seldom develop enough turbulence to lift particles larger than medium-grained sand from their beds. Coarser material may, however, be pushed or rolled along the bottom by the swirling currents of the boundary zone, thus becoming part of what is called the *bed load*.

Except when stream velocities are high, the grains of the bed load do not move continuously but progress downstream by many stops and starts. Consequently, the bed load usually moves much more slowly than the suspended load. The motion of the bed load has been observed through windows in the walls of wooden flumes. Most of the grains roll, though some slide along the bottom, and others bounce along or vault into suspension.

As the velocity of a stream flowing on a bed of sand is gradually increased, the motion of particles on the bed progresses through:

- Short transport of a few individual sand grains,
- Spasmodic movement and deposition of groups of grains,
- Smooth, general transport of many grains.

Under some conditions the scour and deposition of grains produces ripples on the bed. At high velocities large numbers of grains may go into temporary suspension, proceeding as such dense clouds that a distinction cannot be made between the immobile part of the bed and the moving, partially suspended load.

A comparable situation is found in natural rivers, where, except at low velocities, the bed and suspended load apparently grade into one another.

The Dissolved Load

A third way that streams carry material is in solution. The dissolved particles are evenly distributed in the water by diffusion. Here transport does not depend in any way on the nature of stream flow. The chief source of the dissolved load is from inflow of ground water that has percolated slowly through weathering mantle.

Probably very little is dissolved from the channel walls except where streams flow on limestone. Thousands of chemical analyses show that few rivers carry more than 1,000 parts per million (0.1%) of dissolved materials. A general average for many American rivers is about 200 parts per million. It should be noted, though, that this may sometimes form a fourth or a third of a stream's total load.

Load Competence

A stream moving at a given velocity exerts a corresponding force upon the particles of its bed. For particles of the suspended load, the upward-moving currents of turbulence exert forces that depend on their velocities. The total force a current exerts against a grain depends on the grain's surface area, whereas the grain's weight is a measure of its resistance to movement (inertia). Since the ratio of the surface area to weight increases with decreasing size, small grains are more easily moved than large ones. Thus for any stream, the current Will suffice to move particles only up to a certain limiting size. This maximum grain size gives us a simple measure of the stream's transporting power or *competence*.

Considerable experimental work indicates that the diameters of sand grains and coarser debris that can be moved by a stream increase approximately as the square of the velocity. That is, if the velocity is doubled, the diameter of grains that the stream can move is increased four-fold; if the velocity is tripled, the diameters increase nine-fold, and so forth.

The high velocities of steep streams, especially when in flood, make their currents almost irresistible. Some move boulders over 10 feet in diameter. When the St. Francis concrete dam in southern California broke in 1928 the great mass of water suddenly released carried blocks of concrete that weighed as much as 10,000 tons (63 × 54 × 30 feet) for half a mile downstream.

THE NATURE OF FLOW IN STEEP STREAMS

The flow of steep, tributary streams differs markedly from the even or tranquil flow of broad trunk rivers. Not only are their average gradients and hence their average velocities greater, but both gradient and velocity fluctuate more. Changes in gradient of the stream bed create what hydraulic engineers call *nonuniform* flow.

A typical case of nonuniform flow. Water from A passes over the uniform, gentle gradient AB with a moderate velocity. At B the steeper gradient sharply increases the velocity, and hence the stream shallows from B to C. In the stretch from C to D the gradient and, therefore, the velocity are again moderate, and the stream deepens. The alternating rapids, pools, riffles, and waterfalls of most mountain streams illustrate such variations in depth and velocity.

The flow in the A to B stretch of the figure may be considered typical of all streams with low-to-moderate gradients. Turbulence disturbs the surface with mild boiling, rolling, or whirlpool motions. Surface waves produced by impact of the current against the bank or bed—or by dropping a stone into the stream—travel along the surface in all directions, disturbing the general downstream flow of the water. This kind of stream flow, because of its unhurried, though somewhat disturbed, nature, is called *tranquil* or *streaming* flow.

Where such a stream suddenly quickens on a steeper slope its appearance changes abruptly. It breaks away from the surface of the tranquil stream in a curving lip, and shoots swiftly down the steeper slope. If the bed is reasonably smooth, the stream surface is smooth and shimmering; where the stream flows over obstacles, there is a smooth roll or a boiling wave, but the wave's disturbance does not move upstream. Turbulent eddies are apparently flattened and attenuated, for disturbances of the surface are greatly reduced. This kind of flow is called *rapid* or *shooting* flow.

The critical velocity separating shooting from streaming flow cannot be simply defined, for it depends on many hydraulic variables. In general, shooting flow will be produced in a stream one foot deep and several feet wide if the slope is greater than about 3°; in larger streams shooting flow develops on lesser gradients. Here the surface of the swift shooting stream must suddenly rise to the level required by the slower velocity.

This abrupt rise, produced wherever the channel gradient sharply decreases, is appropriately called the *hydraulic jump*. The impact of the shooting water against the slower produces a strong rolling vertical eddy or a seething, air-charged mass of very high turbulence. Except for the waterfall, which is, in a sense, an extreme case of the hydraulic jump, the jump is the greatest dissipator of energy in a stream.

EROSION ON STEEP GRADIENTS

What mechanisms, inherent in the nonuniform flow of steep streams, allow them to erode their beds and banks? Whether a stream moves by shooting or streaming flow, most of its eroding power depends on the energy of large fast-whirling eddies. These eddies are produced mainly by impact against, or deflection around, obstacles in the channel, such as boulders on the bed, bedrock projections, gravel bars, shallow depressions, and deep holes.

During floods, swirling currents may be able to free and move jointbounded blocks of bedrock, a process called *hydraulic plucking*. Large nonuniform-flowing streams produce eddies of great competence. Below Great Falls, Maryland, where the Potomac River moves by streaming flow through a gorge, 15-inch boulders have been lifted as high as 60 feet above the channel bottom during floods. The impact of such missiles must effectively erode both bed and banks. In slower streams, swirling currents armed with coarse sand and small pebbles strongly abrade both moving rock fragments and exposed bedrock. Even fine sand may be an effective abrading agent if the currents are swift.

Effects of eddy erosion can be seen along many streams during low-water stages. Among them are rounding of the projecting rock, fresh angular scars where blocks have been torn from channel walls, and cylindrical cavities (potholes) drilled in solid rock by stones caught in holes on the surface but rolled about by the swirling water.

The strong turbulence of the hydraulic jump causes relatively rapid erosion. Erosion at waterfalls illustrates an extreme in the competence of nonuniform flow. Water falling freely is accelerated by gravity, so that at the end of a second its velocity is about 32 feet per second (20 miles per hour).

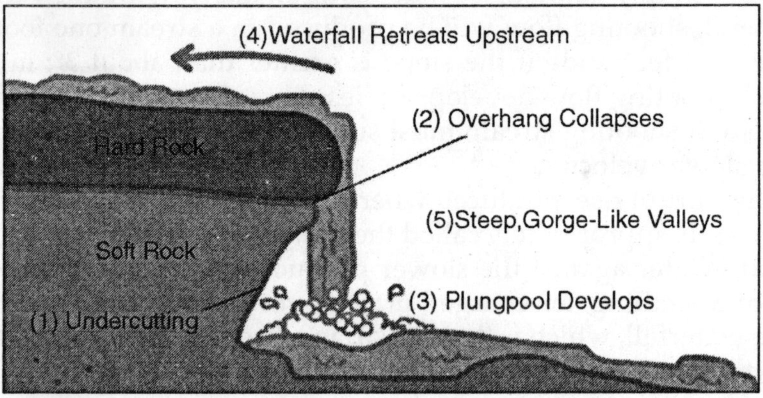

Fig. Erosion at a Small Waterfall

The Niagara River falls at about 50 miles per hour at the base of the 150foot high Niagara Falls. Thus many waterfalls develop deep plunge pools at their bases which are cut back under the falls. Migration of waterfalls upstream by undercutting at the plunge pool and caving of the walls above may contribute greatly to the erosion of a stream valley.

On the Niagara River, for instance, the Niagara Gorge between Queenstown and the Falls, a distance of about 7 miles, was cut principally by headward erosion of the waterfall, not by slow downward erosion along the entire channel.

THE NATURE OF SINUOUS RIVERS

The lower courses of most river systems show features that contrast strongly with the pools, rapids, and waterfalls of steep streams. Trunk rivers commonly flow in broad, smooth valleys sloping so gently seaward as to appear flat. Apparently disdaining the shortest route to the sea, the rivers wind in intricate serpentine courses. The Mississippi River, from Cairo, Illinois, to the mouth of Red River, has a typically sinuous course. Such streams are often called *meandering rivers*, though, strictly speaking, bends called *meanders* should turn through 180° or more. Alternating with sinuous stretches are relatively straight reaches.

The broad flat on which such rivers flow is called the *river floodplain*, for it is more or less flooded when the river overflows its banks. Some floodplains are no wider than the meandering course of the stream—their bends impinge against the valley walls. Many floodplains, however, are far wider than the meanders, and may broaden greatly where joined by sinuous tributaries, as does the Mississippi floodplain, shown by the dotted area.

Low indistinct ridges on floodplains, called *natural levees*, commonly border sinuous river channels. When streams overflow their floodplain, the great decrease in velocity and turbulence of the overbank flow results in deposition of even the finest suspended load. The coarsest part of the suspended load, dropping out just as the floodwaters overtop the banks, builds up the low natural levees. The fine silt and clay are carried farther and deposited on the lowlands of the floodplain, behind the natural levees.

These muddy floodplain accumulations are individually thin, but may build up to considerable total thicknesses over long periods of time. In the lower Nile Valley, silt has risen as high as 15 feet about the ancient Egyptian structures on the floodplain, indicating a depositional rate of 4½ inches per century. The floodplain behind the natural levees may be poorly drained, and lakes, swamps, or indistinct sinuous channels are common.

THE CHANNELS OF SINUOUS STREAMS

Sounding of the Mississippi's bed reveals marked changes of the channel shape from bend to bend. The channel is deepest near the outer bank of each bend. The inner slope is gentle or even convex, and often consists of shifting sand bars.

In the short straight stretch between bends the river shallows considerably, and the channel is more or less symmetrical. These shallow channels between bends, called *crossings*, troubled the old-time river pilots. Some are less than 10 feet deep at low water, as compared to "bendway" depths of 45 or more feet.

This variance in channel shape is directly related to the distribution of velocity and turbulence. In straight reaches or crossings the zone of maximum velocity, bordered by two zones of high turbulence, is midway in the channel. At a bend the conditions are different. The water moves forward into the bend with the direction given by the straight stretch above, so that the zones of maximum velocity and turbulence impinge against the outer bank of the bend. These competent currents contrast strongly with the sluggish flow over the convex slope on the inside of the bend.

EROSION AND DEPOSITION IN SINUOUS CHANNELS

The bank materials are generally unconsolidated sands, silts, and clays; therefore the outer bank of a bend will be easily eroded by the strong currents directed against it. During a flood, when velocities and turbulence are greatest, the channel deepens, causing the outer banks to cave off rapidly. The fine-grained parts of the load thus acquired, the clays and silts, are readily carried away in suspension. Ordinarily, however, the coarser material is not moved far. Experiments show that almost all the sand and gravel is deposited on the next crossing, and over the inside slope of the next bend.

Erosion of the outer bank and deposition on the inner slope of bends cause them to shift in position. This *migration* of the channel, or shifting of the meanders, allows a stream to wander widely over its floodplain. If the floodplain materials are homogeneous, the meanders tend to sweep evenly downstream without lateral shifting. Over a long time the stream reworks the deposits left by previous meander courses and deposits material that is later cut away again as its channel continues to migrate. Thus the coarser sediments of floodplains move only slowly toward the sea.

Erosional and depositional features on many floodplains clearly indicate former positions of sinuous channels, the low stripelike ridges that almost parallel the inside of the bends are beaches or bars left behind as the bends migrated. The crescent-shaped lakes are abandoned river bends. Such *oxbow lakes* develop either by *cutoff* or by *chute* formation. Cutoffs form when one meander is slowed in its downstream migration by a resistant bank, allowing the next bend upstream to catch up with it and cut through the narrowing neck between the two bends.

A chute, on the other hand, forms when a river in flood simply flows over the bar and beach deposits at a bend, and then continues to use this shortcut channel. Chutes sometimes serve as alternate channels around bends, as illustrated in the middle distance of the enlarged map of a part of the Mississippi. Generally, the cutoff or chute channel is more efficient than the channel of the old bend, for, since the distance is shorter, the gradient is steeper, and therefore the velocities and competence are higher.

Velocity and turbulence in the cutoff bends become low or almost nil; consequently, sand is immediately deposited at both ends of the old bend, eventually damming it off from the new channel, and converting it into an oxbow lake.

Significance of Sinuous Channels

Why should streams meander down their floodplains rather than flow in straight channels directly to the sea? In seeking an answer to this question geologists and hydraulic engineers not only have uncovered clues to the cause of meandering, but have come to understand many variables affecting river flow. Observations of the discharge, load, and velocity of natural rivers have contributed much to our knowledge of flow in sinuous channels, but many variables are too intricately interrelated to examine one at a time, and hence their effects on sinuosity are difficult to evaluate.

Because of the variability of natural rivers, investigators are turning more and more to stream models for help in solving both the theoretical and practical problems of stream flow. Simple wooden troughs, used by early experimenters, have been replaced by carefully constructed scale models that not only exactly reproduce channel and valley shapes, but also have measuring and control devices for nearly all the variables of flow. Such models are used considerably in planning major projects of river regulation and construction, and are thus the very intricate "test tubes" of the hydraulic engineer.

Many river model studies of meandering have been made at Vicksburg by the Mississippi River Commission. The models were not mere table-top arrangements, but were of considerable size. In one of the most revealing experiments a straight channel was carefully molded in uniform Mississippi River sand, and water was allowed to run in it for three days.

The stream quickly developed a sinuous course, whose bends swept evenly downstream in the homogeneous, easily eroded floodplain material. The meandering began when the stream, cutting one of its banks more rapidly than it could transport all the debris, deposited a sand bar along that bank.

he sand bar at A forced the stream against the opposite bank, *B*, from which the current was deflected through the points *C, D, E,* and so forth, like a ball oscillating against the sides of an inclined trough. These deflections made the currents at *B, C, D,* and *E* more competent than at *F, G,* and *H*. The banks under lateral attack were eroded, and the load thus acquired was deposited in the first downstream area of low turbulence. Deposits on one bank crowded the stream against the opposite bank, so that the angle of attack on this bank became greater and greater, gradually forming a series of migrating bends.

During the experiments discharge, valley slope, load, and erodibility of bed material were individually varied, in an attempt to ascertain their effects on sinuosity. With increased discharge, the meanders widened in a regular and predictable way, verifying the general rule in nature that big rivers have big bends, little rivers little ones. For each set of flow conditions there was a certain maximum meander width. The chutes and cutoffs are apparently the mechanism—a sort of safety valve—that keeps this neat balance.

Likewise, the valley gradient directly and predictably affected the meander width: the steeper the slope, the wider the bends. A stream with steep gradient thus lengthens its course by forming wide bends, thereby decreasing its gradient. When the streams were not given any load of sand at the head of the model channel, the meanders developed only *after* the stream was able to gain a load from its banks downstream. In experiments with channels molded in partially cemented sand, the streams could gain no load from their banks and did not meander at all.

Thus, as suggested by the mechanism of meander formation, in order for streams to develop sinuous courses, they must have easily erodible banks. This generalization is perhaps the most significant result of the Vicksburg experiments. It is strongly supported in nature by the common association of sinuous streams with wide valleys underlain by loose alluvial deposits.

BRAIDED STREAMS

A limiting case was reached in the experiments when streams were fed more sand than their gradients allowed them to carry away. Here the channel quickly choked with sand bars. The stream was then deflected in all directions into such a spread-out and intricately subdivided cross section that no continued attack could be made on a bank and no meanders developed.

Such streams are called *braided streams*. They are common below glaciers where sediment chokes the stream they also abound in sandy regions where more material is washed to the streams than can be transported in low-water stages. The Platte River in Nebraska is an excellent example.

The Concepts of Base Level and Grade

All streams are limited in their ability to deepen their valleys by the level of the sea—or lake—into which they flow. If this limiting level, called the *base level* of the stream, remains unchanged, the stream tends to stabilize itself on that gradient which is just necessary to transport its load.

Sinuous streams like the lower Mississippi sidecut slowly through the alluvium of their valleys, but alter their gradients very little over long periods of time. Sand and gravel are moved from the outer banks of bends to crossings, and thence to the growing bars on the convex slope of the next bend. Thus over a given stretch of the river, erosion and deposition practically cancel one another's effects, although load is being moved continuously downstream.

Similarly, the average gradient remains the same over considerable stretches of the river, though the bed gradient varies from deep bendways to shallow crossings. If the gradient and average channel shape remain the same, the hydraulic characteristics of the stream—the distribution of velocity and turbulence—do not change, and thus the pattern of erosion and deposition are stable. Stream flow is just adequate to deliver at the lower end of a given stretch a load equivalent to that introduced at the upper end of the stretch.

Streams, or parts of streams, that are thus balanced between erosion and deposition are said to be *graded* or to be flowing at *grade.* If the slope of a graded stretch is reduced—as by slight movements of the earth's crust-the velocity and turbulence of the stream is reduced, and the stream deposits sediment until its channel is built up to the slope required by grade. If the gradient is increased, the velocity and turbulence increase, and the stream erodes its bed to the limiting slope of the graded condition.

Changes in load or discharge similarly force streams out of the graded condition, until a new balance is attained. The steep tributaries of river systems are said to be above *grade* because they are obviously eroding downward into their beds. Rivers flowing *below grade,* characterized by braided courses, tend to deposit their load or *aggrade* their channels.

STREAM DEPOSITS

Besides the alluvial deposits of floodplains, stream deposition may produce deltas, alluvial fans, and extensive alluvial plains. Some floodplains apparently result from long-continued stream erosion. A large number have been formed by the accumulation of stream deposits in pre-existing valleys. The aggradation of most of these valleys seems ultimately to have been caused by crustal warping, changes in sea level, or widespread glaciation.

Erosional Floodplains

Under ideal conditions, floodplains may develop by protracted valley cutting. This ideal process has been summarized by some geologists under the heading "Stages in the Evolution of a Valley." This theory is probably best explained by applying the principles of stream erosion and deposition thus far to a simple hypothetical example. Consider the stream.

As long as it flows on a slope steeper than is required by the graded condition, it will cut into its bed, thereby tending to steepen its valley walls. This steepening initiates downslope movements on the sides of the valley. As long as the stream is competent to remove this debris and, at least intermittently, to attack its bed, its valley will have a V-shaped cross section (stage of youth).

Since the stream cannot cut below its base level, bed erosion gradually reduces the average gradient, thus diminishing its ability to cut down. Eventually the stream will become graded, and, except during the strongest flood flows, its bed will be protected by a veneer of sand or gravel. Though such a stream can no longer cut downward, it may erode its banks by lateral attack against the valley walls.

This will happen if the river is crooked at all. As in meandering streams, the mechanism of lateral erosion is simply the impingement of vigorous, turbulent currents against the outer banks of bends. The weak currents along the inner banks of these bends will simultaneously drop their load as beach or bar deposits. As the stream slowly cuts into its valley walls, it leaves behind a planed-off bedrock surface covered by a thin veneer of sand and gravel (stage of maturity).

Such a process can eventually produce a broad continuous floodplain. The attainment of grade separates the stage of youth from that of maturity.

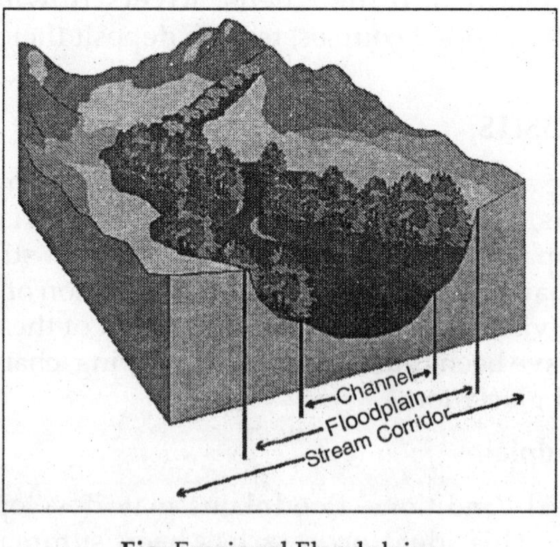

Fig. Erosional Floodplains

This sort of lateral erosion must be much slower than that of streams meandering on alluvium, primarily because of the greater erosional resistance of most bedrock. Another factor is the enormously greater load acquired by lateral cutting and downslope movements. When a stream cuts even a foot into the wall of a deep valley, it makes it possible for downslope movement to add tons of debris to its load.

Because of the great amount of time required to cut erosional floodplains, the probability that some change of conditions in or on the earth will upset the ideal progression of events is correspondingly great. It seems, indeed, that good examples of simple, laterally eroded floodplains are rare. Most present-day streams have been sufficiently affected by crustal movements, sea level changes, or climatic changes to produce more complex valleys. Widespread floodplains seem to be chiefly of aggradational origin.

Aggradational Floodplains

Careful studies of the valleys of many sinuous streams show that their floodplains, instead of being alluvial veneers on a truncated bedrock surface, are underlain by deep deposits of sand, gravel, and mud.

Among these streams are: the Lower Mississippi River; many stretches of the Missouri, Miami, and Ohio rivers; the short sinuous stretches of the lower Colorado, Columbia, and Snake rivers; and the SacramentoSan Joaquin River system of California. It is a notable fact that through the sinuous channels of these streams flows no less than two-thirds of the total runoff of the United States.

The evidence of aggradation is from drill holes and water wells in the floodplains. The Mississippi River Commission used several hundred drillhole and water-well records in its study of the deposits underlying the Lower Mississippi floodplain. These showed that the plain is underlain by alluvial debris ranging in depth from about 100 to over 400 feet. Near Natchez this debris is 260 feet thick, with its base 215 feet below sea level. Drill holes across the present floodplain indicate that the buried surface, the valley of a prehistoric Mississippi, is not a smooth plain but a steep-sided, though shallow, valley with many tributaries.

A detailed study of the Santa Ana River in southern California demonstrates similar aggradation. Here an ancient channel is buried beneath stream deposits 140 feet thick at the present mouth. This filled channel, traced upstream in drill holes, has about the same gradient as that of the modern stream. What has caused the marked filling in of such valleys as these? The deep filling of the Mississippi River Valley was probably induced in part by crustal downwarping (perhaps from isostatic sinking due to the load of the Mississippi Delta), and in part from a greatly increased stream load in the northern tributaries.

The aggradation of the Santa Ana River was probably caused by a rise in sea level since the ancient channel was cut. The ultimate cause of this change in sea level and in the load carried by the Santa Ana and other ancient rivers was apparently the growth and melting of very large glaciers during the recent geologic past. These glaciers robbed the atmosphere, and ultimately the ocean, of enough water so that sea level was distinctly lowered.

As the glaciers melted, not only was this water returned to the ocean, so that the sea level rose, but also large quantities of rock waste released by the glaciers were dumped into the existing river systems. Glaciers, affected the streams of North America in many ways, upsetting the ideal evolution of stream valleys toward erosional floodplains.

Sinuous Streams in Glacial Valleys

Some of the best examples of meandering streams occupy valleys that once contained tongues of moving ice. As noted at the Nisqually glacier of Rainier National Park, the load carried by glaciers is large; streams that flow from the snouts of glaciers commonly have very large loads supplied to them. As the glacier melts away, the load diminishes and the streams stabilize themselves on the flat, debris-filled valley floors. In the process they take on a sinuous course, and slowly rework the fluvio-glacial deposits which they have inherited. Literally thousands of streams in northern America, Europe, and Asia owe their meandering courses directly to glacial erosion and deposition.

Deltas

Any laden stream must drop its load when it enters quiet water such as a lake or sea. Where the Nile emerges from its valley near Cairo, it splits into channels. These further subdivide and flow to the sea on a broad plain of river deposits. Because of its triangular shape, the name *delta* was applied by Herodotus to this flat depositional plain. Deltas may be triangular, often with a convexly curved border against the sea, or irregular with lobelike extensions like that of the Mississippi.

The shapes and sizes of deltas are considerably affected by the wave and tidal characteristics of the water bodies in which they are formed. Thus, the Mississippi and Colorado rivers, which empty into relatively calm and shallow gulfs, have prominent deltas, but the Columbia and Congo rivers have no subaerial deltas at all. The

Columbia's load is distributed by ocean waves and currents for hundreds of miles along the sea coast; the Congo's is somehow washed down a long, deep submarine canyon into the depths of the Atlantic. Where sand-laden streams flow into a deep, still body of water, the layers of deltaic sediment are not simply parallel to the lake or sea bottom, but show a characteristically discordant arrangement. By either bar deflection or overflow on the delta surface the stream tends to seek new channels, often subdividing into several *distributaries* over the delta surface.

The stream deposits formed on the top surface of the delta are called *topset beds* and are commonly thin, because the distributaries must maintain certain minimum gradients across the delta. Thicker deposits, called the *foreset beds,* are laid down on the frontal slope of the delta. The finest part of the load, kept in suspension for a long time by weak currents moving down the frontal slope, is deposited on the lake or sea bottom, in front of the advancing delta. These *bottomset beds* are eventually covered by later foreset layers, often with notable angular discordance.

Most large deltas are much more complex than this ideal example. The Mississippi delta, like that of most other large rivers, shows little or no discordance between topset, foreset, and bottomset beds, probably because the load is fine grained and the sea bottom on which the delta is growing is only slightly inclined.

THE LONG PROFILES OF STREAMS

A fruitful way of finding the degree of balance between erosion and deposition in river systems is to make a graphic study of the stream gradient over long stretches of the stream's course. This is done by plotting the elevations of points on the stream's course against their distances from the mouth, and connecting these points with a line. Such a line is called the longitudinal or long profile of the stream. In order to show variations in gradient it is necessary to exaggerate greatly the vertical scale of such a plot, for the length of streams is commonly very great compared to their vertical fall.

The long profile of the Arkansas River, which rises in the central Rocky Mountains and flows across the Great Plains, joining the Mississippi 440 miles above its mouth, shows slopes consistent with the general description of river systemsr. In the upper course the gradient is steep and irregular; here the river flows in high mountain valleys and, just above Canon City, through the chasmlike Royal Gorge. In contrast, the gradient of the lower 200 miles of the river is even and very low, nowhere exceeding 1 foot per mile. Here the river meanders in a tightly serpentine pattern on a nearly flat, broad floodplain bordered by low bluffs. The notable thing shown by the profile, however, is that between these two extremes the gradient varies regularly and so permits the profile to approximate a smooth concave curve, steepening markedly near the river's origin. Probably at least half of the thousands of long profiles that have been drawn show this general form, though really others show departures from it.

THE PROFILE OF EQUILIBRIUM

Examination of many streams shows that it is especially their sinuous portions on floodplains that have nearly smooth concave profiles. Since these sinuous courses are commonly the effect of a graded condition, it may be said that graded stretches of streams show nearly smooth profiles. The Arkansas River, for instance, shows a fairly smooth concave profile between Pueblo, Colorado, and its confluence with the Mississippi; over nearly all this 1,350-mile course it flows on a floodplain, neither eroding nor aggrading its bed to any marked degree. The profiles of such graded rivers as the Mississippi, the Missouri, the Amazon, the Po, and many others display nearly smooth concave curves everywhere except in their uppermost courses. Nearly smooth concave profiles, then, are the ultimate goal toward which all streams work as they approach grade. They are the commonest type of the *graded profile or the profile of equilibrium*.

The general concavity, itself, of graded profiles raises an important question. If such rivers as the Arkansas are, indeed, graded over certain long stretches, how is it that their gradient changes almost continuously? That is, if a river is at grade on one slope, how can it also be at grade on a steeper one upstream? The factors that apparently allow this are:

- The increase in discharge downstream
- The change in the grain size and quantity of load downstream.

Increase in Discharge Downstream

There is a decrease in the ratio of friction surface to channel cross section as a stream grows. Consider the junction of tributaries. If the channels are all one foot deep with nearly vertical banks, the result of the union would be the removal of four cross-sectional feet of friction sur face. The energy previously dissipated in friction is available for erosion of the channel, resulting in a lowering of the gradient below the junction. Can the actual increases in discharge be correlated with changes in the slope? Though data on the discharge of all the tributaries, particularly the amount of addition from ground water, are not available, a general assessment of the relation between slope and discharge is possible. The Arkansas River is initially formed by the confluence of dozens of mountain streams which are, generally, flowing above grade.

From Pueblo, where the river enters the plains, to Dodge City, contributions are restricted to small intermittent streams and seepage of underground water. In summer much of this stretch is braided. Between Dodge City and Arkansas City the discharge is increased by many small tributaries, and below Arkansas City by three major tributaries: the Cimarron, Neosho, and Canadian rivers. The profile, the gradient flattens below the junction of these tributaries. A similar relation has been noted in the Lower Mississippi River, whose gradient and discharge have been carefully studied. Some statistical studies have been made to determine the cause of the concavity of a stream's profile. For example, the discharges of rivers in the French Alps have been measured regularly, day after day, at 73 places. When the mean maximum discharge at each station was graphically plotted against its local gradient, the points approximated a smooth curve—strongly suggesting a causative relation between slope and discharge.

Effect of the Load

All graded streams do not show such a simple relation, probably because of the varying loads brought to the trunk stream by different tributaries. If the competence or capacity of a tributary is con siderably greater or less than that of the trunk stream, the confluence will force a steepening or flattening of the trunk stream's gradient. Such a situation may be illustrated by an extreme case on the Colorado River of Arizona. Here tributaries are dry through most of the year, but during the seasonal floods their swift currents move great quantities of coarse debris into the main river. Much of this material is too coarse to be moved by the more sluggish Colorado, and so it piles up as an apron of boulders, called a *debris dam*. Since these boulders can be transported only after they have been slowly ground down by abrasion, from the standpoint of human history, debris dams cause practically permanent convexities on the river's profile.

Decrease in Grain Size Downstream

Though the debris dams of the Colorado provide striking examples of the effect of the grain size of the load on a river profile, more important, though more subtle, examples may be found in graded, sinuous rivers. A careful study of the bed materials of the Mississippi River shows a marked decrease in the size of the particles downstream.

The change in grain size of 600 samples that were collected.

The average composition of the delta sediments, which are 70 per cent silt and clay, is further evidence of this general change. On the Rhine River, between Basel and Bingen, there is a marked decrease in pebble size downstream that is closely correlated with a flattening of the river profile. It is noteworthy that the discharge increases only slightly over this stretch. Thus in some rivers the slow attrition of the bed load downstream appears to be a factor in allowing the stream to flow on an ever-decreasing slope.

DEPARTURES FROM THE IDEAL PROFILE

The smoothly concave curve of the ideal profile is closely approached but probably never attained by natural streams. Even in such rivers as the Lower Mississippi, probably as well-graded a stream as could be found, the profile shows slight but abrupt changes of slope between many individual stretches. This is not at all abnormal. For one thing, the inflow of most tributaries demands such changes in gradient. Note that the vertical scale of this profile has been exaggerated 2,100 times in order to show these slight changes. Other streams show much more marked changes in slope that must be considered to be actual departures from the ideal profile.

One of the principal values of the concept of the ideal profile is that marked departures from it call attention to exceptional circumstances. The debris dams on the Colorado are an example. Lava flows, landslides, or drifting sand dunes may upset the graded condition of a stream and impose an irregularity on its profile. Glaciation in the recent geologic past has greatly affected the slopes of many streams. Most, if not all, of the major streams of the northern United States have irregular profiles inherited from glacial erosion and deposition. Earth movements are also a factor, if they do not allow the stream quickly to adjust by degradation or aggradation. Some of the anticlines in the south central part of the state of Washington have apparently been uplifted rapidly and recently enough to outstrip the streams flowing down their flanks. Even in streams that have long been cutting toward grade in a relatively stable region, there may appear distinct changes in slope—called *nick points* —on the profile caused by local outcrops of especially resistant rocks. The profile both above and below such a nick point may show a smoothly concave shape. Considerable time is required for the ideal profile to form in regions of resistant rocks. Therefore all kinds of structural or climatic "accidents" may intervene before the graded profile can be formed. It is a tribute to the eroding power of running water that many large streams flow on grades that approximate the profile of equilibrium.

DOWNSLOPE MOVEMENTS

Land forms result from the interplay between two opposing forces. On the one hand, earth movements and depositional processes raise certain parts of the crust; on the other, erosional forces constantly work to level them. Streams and downslope movements, attacking uplifted areas, produce land forms whose shapes depend chiefly on the structure of the underlying rocks, the climate, and the nature of the uplift that initiated erosion. The first two of these factors can ordinarily be determined, and, if the processes of stream erosion are well understood, it is often possible to interpret the nature of the constructional force.

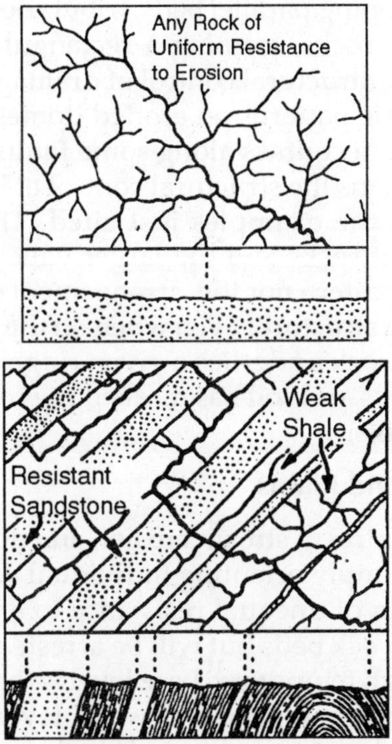

Fig. Structural Control of Stream Patterns; left, Dendritic Pattern; Right, Trellis Pattern

HEADWARD EROSION AND LANDSCAPE PATTERNS

Erosion in the steep headwater tributaries of a stream system not only deepens its valleys but also extends them by *headward erosion.* Stream systems thus tend to expand as they reduce the land, and, coupled with downslope movements, create new ridge and valley systems or modify old ones. Headward erosion is guided by the arrangement of slopes and the structure of the rocks being dissected.

These two factors thus control the resulting pattern of stream courses and ridge lines. If the rock and its mantle offer uniform resistance to erosion, and if the slopes of the land surface are more or less randomly arranged, tributaries will branch and erode headward in a random fashion, producing a *dendritic pattern,* resembling the branching limbs of a tree.

If the rocks being dissected are of unequal resistance, the extension and downcutting of tributaries will be most rapid on the weaker rocks, which will thus come to underlie valleys or lowlands between ridges or uplands of resistant rocks. Thus in a stream system flowing across a region of steeply dipping parallel beds, tributaries tend to concentrate on the belts of weak rocks, forming a rectangular or trellis pattern. Other examples of the structural control of drainage patterns and ridge lines are the concentric patterns on eroded domes and basins and the strikingly linear stream courses along some faults.

In temperate regions the structural control of land forms is usually more subtle than in the examples just cited. The major ridges are underlain by resistant rocks and the larger valleys by soft rocks. Yet neither the drainage pattern nor the, arrangement of slopes gives much clue to the underlying structure. Even when on the ground, it is difficult to tell what resistant rock underlies a given ridge, because downslope creep of a layer of grass and soil has masked the rock contacts, forming smoothly rounded bills.

Structural Terraces and Plains

The structural control of stream erosion and downslope movement can go further than simply orienting individual valleys and ridges. In thick horizontal strata of unequal resistance to erosion, a stream will cut quickly through weak beds but will be arrested by a resistant layer. Thus, a general, though temporary, base level is formed, and headward erosion by tributaries may strip the weak layer over a large area, forming a *structural plain* on the resistant bed.

In an arid or semi-arid climate the dissected upland border is eroded back as a steep embayed slope, and temporary upland remnants called *mesas* and *buttes* are commonly left behind to attest to the stripping of the plain.

If a stream cuts its valley in a sequence of alternating weak and resistant beds, each resistant bed will form a step or *structural terrace* in the side slope of the valley. Such terraces are common in the Grand

Canyon of the Colorado and where the Columbia and Snake rivers cut into the horizontal lavas of the Columbia Plateau.

STREAM TERRACES

More common than structural terraces are valley terraces that cannot be explained by differences in the resistance of flat-lying strata. Some such terraces are paired across a valley, others are single. Some are made up entirely of river deposits, others are cut in bedrock but have veneers, channel fillings, or residual patches of river gravels on their flat surfaces. From such facts as these, it seems certain that these terraces are the remnants of old floodplains, now incised by the streams that once made them. Why a stream thus dissects its floodplain is in many cases difficult to understand. In general, terraces are initiated by earth movements, changes in sea level, or any factor that puts the stream above grade. Terraces obviously caused by earth movement were formed shortly after the Pleasant Valley earthquake of 1915. The 12-foot offset along the front of Sonoma Range produced a sharp fall in the profile of streams crossing the fault. Erosion at this step was rapid, so that by 1930 it had retreated far upstream, leaving a 12-foot channel that broke the old floodplain into a pair of matched terraces.

Because this stream now sidecuts at its new graded level, the terraces will be reduced in width, though probably parts of them, particularly downstream from projecting ridge spurs, will remain for a very long time. It is of interest to note that even if the 1915 fault offset had not been observed, a clue to the movement could be found by examining the stream terraces, for they parallel the present stream profile in the Sonoma block, but end abruptly at the fault line.

Another example of terraces caused by faulting occurs on the west side of the Panamint Mountains, California. After faulting formed a series of small scarps across the large alluvial fans bordering the range, streams cut trenches in the uplifted portions of the fans and formed new fans below the scarps. Since, stream valleys within the mountains had no floodplains, the terraces extend only from tile new fault line to the mountain front.

Extensive Erosion Surfaces of Low Relief

An example of a well-marked erosion surface of fairly low relief, now being destroyed by erosion, but too extensive to be called a terrace, occurs in the central and southern Appalachian Mountains. This old surface is well developed in the vicinity of Harrisburg, Pennsylvania, and so is called the "Harrisburg surface". In some areas at or near the east edge of the range this surface is many miles wide, undulating, and mostly underlaid by deep residual soils. The surface has been formed on all but the most resistant rocks of the region, cutting across flat and tilted beds alike. The Susquehanna and other rivers have cut 200- to 300-foot gorges into it, and lesser streams have partially dissected it. The Harrisburg surface exists only as valley terraces in the parts of the Appalachians that are dominated by ridges of resistant sandstone. The terraces are 200 to 300 feet above the streams in their lower courses, but the terrace gradients are commonly less steep than those of the present streams, so that upstream the terraces decrease in height. Most river valleys in the southern Appalachians have similar terraces that can be approximately correlated with the Harrisburg surface. In some places broad passes between drainage systems, followed by roads and railways, are on the Harrisburg surface, though it is of interest to note that locally the surface is not at the same elevation in different valleys on the two sides of such passes.

Widespread erosion surfaces, commonly called *peneplains,* similar to the Harrisburg surface but generally with still lower ridges, occur in many parts of the world. Some, as in the central part of the United States, from Missouri and Kansas south, lie near the graded level of through-going streams. How are such extensive, nearly flat surfaces formed? Careful study of the better preserved surfaces reveals that many are the result of erosion by streams flowing very near their base level. Evidence for such an origin lies in the presence of a deep zone of thorough weathering, the fact that all but the hardest rocks-in all attitudes—are truncated, the presence of residual stream channels and stream gravels, and the complete absence of surficial marine or glacial deposits. Though these surfaces are similar to terraces in some respects, they differ from them in other ways besides their size. Floodplains may be formed by the sidecutting of a single stream, but surfaces such as that at Harrisburg are produced by downslope movements combined with the work of many streams. For this reason these surfaces are not flat, but undulating, their low crests and gentle slopes being the last remnants of hills and ridges. Though it is tempting to suggest that such extensive plains are to be expected as the common end product of subaerial erosion, the present status of the earth's landscapes does not bear this out. All such surfaces of low relief that have been found by careful study to be the result of stream erosion are, like that at Harrisburg, variously uplifted, warped, dissected, or otherwise imperfect. However, the existent remnants of these surfaces are evidence of more stable periods preceding the present, and, like stream terraces, locally enable us to estimate the nature of crustal and climatic changes during the relatively recent geologic past.

Chapter 8

Glaciers and Glaciation

Glaciers are slow-flowing, thick masses of ice. *Snowfields* are thinner and essentially motionless masses of permanent snow. Snowfields and glaciers are savings banks in the water economy of the earth. Each year, some of the water evaporated from the seas and lands falls as snow. Most of the snow melts in summer, but in polar regions and at high altitudes part of it remains throughout the year in glaciers and snowfields. If withdrawals by melting and evaporation exceed deposits as snowfall, the balance in the bank decreases and the glacier recedes, but if precipitation exceeds withdrawals, the glacier grows and spreads outward.

THE SNOWLINE

The lowest altitude at which permanent snow rests is called the *snowline*. Obviously, it does not persist always in one position but is affected by variations in the annual snowfall, temperature, wind direction (which controls drifting), and topography (which controls both avalanching—snowsliding —and shading from the sun).

Mean annual temperature decreases at higher altitudes and at higher latitudes, hence the snowline is higher near the equator than in polar regions. The amount of snowfall also affects its position. Thus, the snowline stands over 5,000 feet higher (8,000 to 9,000 feet) on the dry eastern side of the St. Elias Mountains along the Alaska-Yukon border than on the wet western side (2,500 to 3,000 feet), and it is lower in well-watered Norway than on the colder but dry Taimyr peninsula.

In much of Siberia, northern Alaska, and Canada, the mean annual temperature is low enough, but the precipitation is too scanty to support permanent snowfields and glaciers. In such areas, the pore moisture in the soil and rock is frozen to considerable depths, forming a great sheet of *permanently frozen ground*.

SNOWFIELDS

Above the snowline, on all except the steepest and windiest slopes, lie permanent snowfields. Excavations show that the beautiful geometric patterns of new-fallen snowflakes do not persist at depth. Instead, the snowfield consists largely of small rounded granules of ice about the size of birdshot. This material, called *firn,* has been formed by compaction of the feathery snowflakes and by alternate thawing and refreezing of their edges.

This melting and refreezing is not due entirely to variations ill air temperature above the snowbank. Water expands 9 per cent when it freezes; hence, pressure tends to lower the melting point of ice. Thus, even at temperatures slightly below the normal freezing point (0° Centigrade or 32° Farenheit), the snowflakes deep in a snowbank may be so tightly packed by the weight of later snowfalls that a thin film of water forms at the points of contact between them.

This water film may then move to a nearby pore space, where the pressure is lower, and refreeze into ice, even though there has been no change of temperature. Rains falling on the snowbank and meltwater formed on a warm day are absorbed as in a blotter by the granular firn. Perhaps the following night this water freezes, assisting in the compaction of the firn by excluding air spaces.

Thus, deep in the snowfield the firn gradually changes to interlocking granules of solid ice. Crevasses in glaciers commonly show all steps in the transformation: Snowflakes '! Firn '! Solid Ice.

GLACIERS

If snow and firn merely continued to accumulate each winter, even the highest mountains in arctic regions would eventually be buried. Ice, however, is not a strong mineral, and its strength limits the height to which it can accumulate. Before any great thickness is reached, the ice at the bottom begins to creep and flow outward under the overlying weight. Thus, from the snow and firn a *flowing ice mass,* or *glacier,* is formed.

GLACIER MOTION

Glaciers move so slowly that the motion cannot be seen, but the fact that the ice actually is moving can be readily proved by driving a straight row of stakes across the surface of the glacier. In a few days or months the straight line becomes a curve, bent downstream in the centre of the glacier where the speed is greatest.

The rate varies in different glaciers and also with the season of the year. In a few of the Alaska and Greenland glaciers, velocities of more than 150 feet per day have been recorded, but such rates are exceptionally high. The lower parts of some glaciers on the coast of Alaska appear to be practically stagnant.

Indirect proof of glacier flow is given by the rocks frozen into the glacier. Most large glaciers contain abundant boulders which must have been carried from far upstream by the moving ice, because no similar rock is found in the valley walls nearby. The bodies of climbers who crashed to their death in the treacherous crevasses of the Bossons glacier in the Alps were released 41 years later at the terminus of the glacier several miles below.

Nature of Glacier Flow

The upper surface of glaciers is riven by crevasses, and the ice appears to be hard and brittle. It shatters like glass under a sharp blow, just as does an ice cube from the refrigerator. It is difficult to imagine such a substance flowing. Actually, there is little or no evidence that the brittle, upper part of a glacier does flow. The abundant crevasses prove that fracture is the chief form of yielding. Deep within the glacier, however, deformation is almost entirely by flowage. Glaciers are perfect examples of the effect of scale on strength. Ice in small masses is rigid and brittle, but in glacial masses it flows under its own weight. The brittle, crevassed surface ice is rafted along on the flowing ice below.

The mechanism of glacier flow is complex. Microscopic studies of ice obtained from glaciers show that some of the ice crystals have been bent, others have glided along the sheets of atoms parallel to the base of the hexagonal crystals, and still others have been granulated and sheared. Furthermore, the crystals tend to be arranged in parallel planes, as in foliated metamorphic rocks, suggesting that they recrystallized during or after movement.

Such recrystallization is demonstrated by the coarser crystals of glaciers as compared to those in snow and firn. Crystals in the firn of Alpine snowfields average less than ¼ inch in diameter, but at the glacial snout the grains average about an inch in diameter. Crystals 6 to 8 inches in diameter may be collected from the edge of the Malaspina glacier in Alaska. The large crystals must have grown by recrystallization of smaller ones.

The ice at the front of a glacier, or in deep glacial crevasses, is generally layered. The layering superficially resembles stratification. It lies roughly parallel with the floor of the glacier and curves upward along its walls. Measurements at the glacier front have shown that the ice in adjacent layers moves at different speeds, proving that the layers are separated by shear surfaces. This layering is one manifestation of shear flow—the layers are due to the development of many closely spaced surfaces of shear within the glacier along which the ice and contained rock detritus are dragged forward and streaked out by movement.

The shear-banding of glaciers closely resembles the foliation of many gneisses and schists. Clearly, ice under a relatively small load is no longer brittle, but reacts to differential pressure by flowage and recrystallization. We have, in glacier ice, an example of a metamorphic rock, completely transformed from the original snow and firn into a wholly different streaky crystalline mass by movement and recrystallization under pressure.

KINDS OF GLACIERS

The form of a glacier is largely controlled by the topography over which it flows. *Valley glaciers* are streams of ice that flow down the steep-walled valleys of mountain ranges. Fed by snowfields on slopes and in catchment basins above the snowline, the glacier may extend far below the limit of perpetual snow. All glaciers end at the point where the rate of melting at the front equals the rate of ice replenishment by flowage.

Valley glaciers are typical of lofty mountain ranges the world over. Most of those in the United States, except for the extensive glacier system on Mount Rainier, Washington, are short blunt ice streams only a few hundred feet thick. It is difficult to distinguish many of them from snowfields; indeed there are all gradations.

The Rocky Mountains, the Cascade Range, and the higher parts of the Sierra Nevada contain hundreds of small, rounded or irregular ice masses that lie in steep-walled clefts or else occupy amphitheater-like depressions opening out over steep slopes or cliffs. Such masses are called *cliff glaciers* or *hanging glaciers*. At the foot of the St. Elias Mountains in Alaska several of these long systems of valley glaciers emerge from the mountain front and spread out on the plain below to form the Malaspina glacier, a lobate mass of ice covering 800 square miles. Such glaciers are called *piedmont glaciers*.

The largest glaciers are huge ice sheets, called *continental glaciers*. They are found today only in the polar regions. All of interior Greenland (about 637,000 square miles) is covered by ice, leaving only a narrow fringe of land along the coast.

The Greenland glacier spreads outward in all directions from two high points in the interior. Locally, the coast is bordered by lofty mountains through which the glacier spills, splitting up between the peaks into valley glaciers which then descend the coastal valleys to the sea.

Antarctica supports a much larger continental glacier, estimated to cover 5,000,000 square miles—an area larger than that of the United States and Mexico. The Antarctic glacier overrides the coast and projects into the "shelf ice" formed by the freezing of the sea. In places, the Antarctic glacier is held back by mountains through which the ice escapes in a series of large valley glaciers.

The famous Beardmore glacier, ascended by several of the early explorers in their quest for the South Pole, is one of these outlet rivers of ice from the high polar ice mass. The Beardmore glacier is 300 miles long, 12 miles wide, and extends far out into the shelf ice even after it reaches the sea. Small masses of radially spreading ice are found on Iceland, Spitzbergen, in parts of Scandinavia, and on the islands north of Canada.

They are called *ice caps* to distinguish them from the huge continental glaciers. Glaciers are much more powerful agents of erosion than rivers, but, because ice is restricted to high altitudes and to polar regions, running water moves far more rock debris each year than does ice.

ACQUIRING OF LOAD

Frost Weathering and Avalanching

Glaciers acquire rock debris in several ways. Valley glaciers commonly cover only a small part of the mountains in which they lie. On the gentle slopes between and above them lie extensive snowfields, but on steep windswept slopes even snow is unable to cling throughout the year. Hence, great expanses of craggy ridges, peaks, and cliffs rise above the glaciers. These bare rock surfaces are strongly shattered by frost action.

During the daytime, meltwater from snowbanks seeps into the rock crevices, and then, perhaps that same night, freezes and breaks blocks loose from their parent ledges. Small grains, loose chips, and even great blocks of rock thus freed tumble down the steep slopes and accumulate in talus piles on the edge of the glacier. Much debris is also carried down by avalanches (snowslides) during the spring thaw. Freshets from melting snow or summer rains also wash much debris onto valley glaciers.

Gradually, by being buried under new accumulations of snow or avalanche deposits, and by falling into crevasses opened by the motion of the glacier, much of this rock debris becomes incorporated within the ice, especially along its edges. The edge of a valley glacier may contain so much debris that it shows up in photographs as a conspicuous dark streak.

These stripes of dirty ice and loose rock along the edges of the glacier are called *lateral moraines*. Where two valley glaciers unite, the adjacent lateral moraines merge and form a *medial moraine*. If tributary glaciers are numerous, many medial moraines may streak the surface of the main glaciers. The moraines are not just surface features, they are wedgelike masses that extend deep into the glacier.

Meltwater Shattering

The exposed crags above a glacier are subjected to strong frost action but a glacier blankets its floor from extreme temperature changes. However, the rock floor of a thin glacier may be exposed to meltwater at the bottom of crevasses, and here a change in temperature will cause some frost shattering.

At the head of most valley glaciers is a deep arcuate crevasse, or series of closely spaced crevasses, called the *bergschrund*. The bergschrund gapes open in summer, but is filled or bridged over by snow in winter. It is inferred that the bergscbrund is formed by downstream movement of the plastic ice in the lower part of the glacier which breaks the brittle ice above and pulls it away from the rock wall.

Adventurous observers have descended into bergschrunds on ropes and found the lower parts to have a rocky wall on the upstream side and an ice wall on the other. The rock wall is shattered and riven by joints, some of which gape open and disclose slight displacements of the angular blocks of rock that they outline. Some of the blocks are free but have moved only slightly; others lean out against the ice in precarious unbalance; and the rest, entirely free from their parent ledges, are now incorporated in the glacier.

During the day, water from the melting snow pours over the face of the bergscbrund, filling the joints. The almost nightly frost loosens new blocks. Obviously, here is a zone of active glacial erosion, sapping the headwall by the plucking of frost-rifted blocks, steepening the cliffs behind, and gradually extending the glacial channel headward. An open crevasse, however, cannot extend to depths of more than about 200 feet.

Experimental tests on ice under pressure show that at loads corresponding to greater depths ice will flow plastically and slowly close the deeper openings. Nevertheless, certain typical features of glacier heads indicate that shattering somehow does go on to greater depths.

One factor is that the meltwater itself opens channels in the ice at the bottom of the bergșcbrund. On hot summer days much meltwater cascades into the bergschrund. A "chinook" wind or a warm summer rain may greatly increase the volume. Despite this great volume of meltwater, the bergschrund practically never fills with water to the point where streams emerge and flow over the ice. Apparently, the inpouring water melts its way down between the rock face and the glacier far below the bottom of the bergschrund, possibly, indeed, to the very base of the glacier.

Actually, local melting of the deep ice should be expected from physical principles; the compressed ice at depth melts at a slightly lower temperature than the ice and snow at the surface. Meltwater pouring into the crevasse also carries heat downward to melt more ice deep within the glacier. Plastic flow of the ice tends to close the meltwater tubes, but as long as heat is being continuously transferred downward, melting of the walls may offset the constriction by flowage. Furthermore, water in the tubes may be under high hydrostatic pressure. and the pressure opposes plastic closing.

Thus, we have a mechanism by which the rock wall at the head of a glacier may be wetted periodically to great depths. But, just as periodically, freezing cuts off the supply of water and the accumulated meltwater within the glacier and in the cracks in the rock wall may ultimately freeze.

As a result, the rock wall is shattered and disintegrated to depths much greater than the bottoms of the crevasses. Frozen meltwater also adds volume to the glacier. Such a glacier, now frozen tight to the headwall, moves slowly forward, drawing out the shattered blocks of rock with it, and tending to open new cracks and crevasses to be filled by a further flow of meltwater.

It must be remembered, however, that meltwater in the deeper part of the glacier is protected from day and night changes in temperature. Deep within a glacier the temperature remains almost constant at the freezing point, and ice and water are in physical equilibrium with one another. Change from liquid to solid may occur here chiefly in response to variations in pressure.

A crack opened by the pull of the downsliding glacier will almost immediately fill with ice—the lowered pressure within the crack causes any moisture gaining access to it to freeze. But, in freezing, a gram of water gives up 80 calories of heat, and, if this heat is not dissipated, freezing will stop. For these reasons, meltwater shattering must be less effective at great depths than on the wall of the bergschrund.

Cirques

Downstream flowage of the glacier removes the shattered debris and exposes new rock surfaces to attack. Where valley glaciers have melted away completely, exposing their headwalls, the glaciated valley generally ends in a semicircle of high cliffs that bound an arcuate rock basin generally occupied by a beautiful mountain lake. Such cliffed valley heads are called *cirques*.

The curving cliffs at the head of the cirque are shattered clear to their bases and they generally meet the smoothed and polished floor of the valley at a sharp angle. The jagged and shattered cliff face—the product of frost action and meltwater shattering-contrasts strikingly with the smooth valley floor—the effect of rasping by glacial ice.

THE GLACIAL RASP

The head of a valley glacier is a zone of rapid accumulation of snow and ice heavily charged with rock debris. If accumulation has been rapid enough to produce a steep downstream gradient of the top of the glacier, outward movement in a thin ice mass may occur along curving shear planes similar to the movement in a landslide.

In a thicker and longer glacier, extrusion flow of plastic material may be less concentrated along shears, but it will follow similar paths. As the ice moves across the bedrock floor, rock debris frozen in its base is dragged across the glacier bed, abrading the bedrock. The glacier thus becomes an extremely effective cutting and polishing tool; it acts like a gigantic rasp, gouging into and scraping off irregularities on its bed, and smoothing and grooving the rock beneath it.

Debris in Continental Glaciers

Practically the entire landmass occupied by a continental glacier is covered with ice; thus frost shattering and talus accumulation can take place only around relatively rare islands of rock that project through it. Also, little debris can be won by meltwater shattering of the rock floor at the bottom of crevasses. Yet, the ice at the margins of the Greenland glacier and other ice caps is heavily charged with rock debris, just as in valley glaciers.

Some of this debris is the eroded soil that formerly covered the land over which the glacier flowed. In many localities, where continental glaciers have melted back along their edges, a striking feature of the rock floor formerly covered by ice is the almost total absence of residual soils and of preglacial unconsolidated mantle such as stream gravels and floodplain deposits.

The soil and loose rock have been frozen into the lower part of the glacier and dragged off bodily by the moving ice.

As in large valley glaciers, rock floors under continental glaciers are smoothed and polished. On these polished floors are numerous scratches and even deep grooves. Obviously, the glacier has gouged and abraded material from its bed.

Rate of Erosion by Glaciers

The rapidity with which a glacier abrades its bed depends on four factors: the resistance to abrasion and plucking of the particular bedrock, the abundance of cutting tools (rock fragments) frozen into the base of the ice, the speed of flow, and the weight (thickness) of the ice.

Thus, thick continental glaciers flowing over weak or shattered rock are powerful agents of erosion. Similarly, on the floor of a cirque, glacial abrasion is generally great because here the ice is likely to be thickest and it is abundantly supplied with cutting tools.

DEBRIS RELEASED BY GLACIERS

Till. That glaciers are effective eroding agents is well shown by the vast amount of debris released at the glacier front by melting. Piled in hummocky ridges (*moraines*) along the front of the ice are heaps of boulders, sand, and silt, heterogeneously mixed without appreciable sorting or stratification. Such unsorted debris, deposited directly by the ice, is called *till*. The proportion of boulders to fine material in till ranges widely. Some till is largely coarse boulders, but that from thin ice caps eroding shale and limestone may be chiefly clay and silt, with only scattered boulders.

Rock fragments found in till are not shaped like those in stream or marine deposits. Stream and beach pebbles are commonly rounded, but most of the fragments in till are subrounded or sharply angular. Some have been cracked and broken by the crushing weight of the overriding glacier; many, especially in valley glaciers, are joint blocks loosened by frost action and little modified during transport.

Some fragments show one or more nearly flat, grooved, and polished surfaces. Obviously, these flat surfaces were faceted and smoothed as the boulder was dragged against the bedrock floor, or against other rock debris within the ice.

Rock Flour

The effectiveness of the glacial rasp is especially obvious from the large amount of fine material (silt and fine sand) released at the glacier front. Turbulent streams of milky water generally cascade from one or more tunnels in the front of a glacier. If we collect a beaker of the roily water and let it stand, most of the suspended material promptly settles to form a thick layer of silt and fine sand on the bottom of the beaker. A very small amount of fine clay particles may remain suspended for hours or days.

The silt, examined with a microscope, differs strikingly from the mud brought down by freshets in an unglaciated region. The minerals composing the silt of the glacial streams are not the ordinary soil minerals characteristic of chemical weathering. Instead, small sparkling cleavage fragments of unweathered feldspar and other undecomposed minerals are abundant, and there is an almost complete absence of the yellow iron stain, the black and gray colors from humus, and the slimy clays so characteristic of chemical weathering.

Clearly, most of this material is not an ancient soil removed by the glacier, but is *rock flour* made by the grinding and crushing of rock fragments against one another and against the bedrock.

Glaciers release an amazing quantity of rock flour. Glacial streams are almost invariably overloaded. They quickly deposit the coarsest material (rubble and coarse sand), building up a sloping alluvial plain over which they spread in braided courses. Most of the rock flour, however, is carried far beyond the front of the glacier and builds floodplain deposits farther downstream, or is spread into lakes as extensive delta and bottom deposits. Such rock flour forms great *terraces of white silt* along the rivers of British Columbia and Alaska.

The powdered rock is also widely scattered by the winds. Extensive unstratified deposits of *loess* (a wind-deposited soil composed chiefly of undecomposed silt particles) are found just beyond the limit of glaciation in many areas. Loess characteristically, contains abundant small tubelike holes where grass roots have grown.

FORMS OF GLACIAL DEPOSITS

The debris dumped by glaciers and by the streams flowing from them is called *glacial drift*. The drift assumes characteristic and easily recognized topographic forms.

Moraines

Unstratified glacial drift (*till*) is deposited directly by the ice and occurs chiefly in the form of moraines. The term *moraine* is used both for material deposited by the ice and also for the material upon or within the moving ice. The largest and best developed moraines are generally at the front of the glacier.

If the ice front is in equilibrium, so that the rate of flow equals the rate of melting, it will remain essentially fixed in position for a long time. Here, the debris carried by the ice is released and accumulates in great hummocky morainal ridges. Because the ice front fluctuates with mild climatic changes, however, more than one morainal ridge is likely to form. The moraine that marks the farthest advance of the glacier is called the *terminal moraine*. Moraines that mark stages of halt as the glacier recedes are called *recessional moraines*.

Most terminal and recessional moraines of continental glaciers are broadly Iobate in plan and can be followed, with minor interruptions, for miles. The moraine at the front of a valley glacier is generally a high crescentic ridge curving around the front of the valley glacier and extending up its sides as lateral moraines. Small recessional moraines of valley glaciers may be almost buried in outwash debris.

During the advance and withdrawal of a continental glacier some till is deposited beneath the ice and overriden by it. Till may also be scattered over a glaciated area by melting out of the englacial (ice-enclosed) debris as the ice recedes. This irregularly scattered till, called *ground moraine*, is not concentrated into definite ridges like the terminal moraine.

Some is found packed into hollows such as preglacial stream canyons; some is plastered around and upon low hills of the bedrock floor. Actually, the ground moraine is the most widespread deposit of a continental ice sheet. In most places it is spotty and only a few feet thick, but near the margin of the ice sheet it may cover hundreds of square miles in an almost continuous veneer.

In some areas, the ground moraine has been moulded into clusters of closely spaced hills, each shaped like the bowl of an inverted spoon, with the steeper slope facing the source of the ice. These streamlined hills are called *drumlins*. They vary in size, but are commonly about a mile long, ¼ mile wide, and 50 to 150 feet high. Drumlins occur in clusters of scores or hundreds. Each drumlin in the cluster has its long axis roughly parallel to the direction of flow of the former glacier.

Although we cannot see drumlins forming in modern glaciers, they are doubtless masses of subglacial debris overriden and moulded into streamlined forms by an actively flowing ice sheet. Excavations show they are composed largely of till, commonly a sticky variety containing much clay. Some drumlins have cores of bedrock; others are composed entirely of glacial debris.

Stratified Glacial Drift

Meltwater from the glacier carries off much of the finer debris and deposits it either on the beds of the overloaded streams (*glacio-fluvial deposits*) or in lakes (*glacio-lacustrine deposits*) and in seas. In places, the deposit may consist of *pitted outwash*, characterized by numerous undrained depressions ranging in size from mere dimples a few feet wide up to great holes a mile long and more than 100 feet deep. These holes are called *kettles*. They are formed by the melting of ice blocks that were detached from the main glacier as it wasted away, and were then partially or wholly buried in stream-deposited glacial debris.

The streams emerging from valley glaciers generally have somewhat steeper gradients than those from continental ice sheets. Nevertheless, they are so greatly overloaded that they immediately aggrade the valley floor to form a long narrow floodplain of coarse gravel and sand which grades downstream into finer sands and eventually into wide floodplains of silt. These aggraded valley bottoms, floored by glacio-fluvial debris, are called *valley trains*. When the glacier melts completely, the stream, being no longer overloaded, generally erodes the valley train into a series of terraces.

Stratified sand and gravel, containing here and there an irregular patch of till, forms long winding ridges in some areas from which continental glaciers have melted. Most of them are less than 100 feet high and a few hundred feet wide, but they may be several miles or even ten of miles long. They are called eskers, a term derived from the Gaelic meaning "ridge."

Eskers appear to be the deposits of aggrading streams that flowed in tunnels beneath the ice, or along the bottoms of crevasses, for some of them merge downstream into fans of outwash or into deltas. Most eskers must have formed when the ice sheet had wasted to a thin, almost stagnant mass, for, obviously, the strong thrust of an active glacier would close the tunnels and crevasses and scatter the esker deposit into ground moraine.

Where a glacier-fed stream enters a lake, a *delta* forms which may grow until it converts the lake to a swampy plain. The delta material is typically cross bedded and may range in grain size from coarse gravel to fine silt.

On the floors of glacial lakes, mud and silt accumulate as thinly stratified layers. Each layer is commonly only a fraction of an inch thick, giving the deposit a conspicuous paperlike stratification.

Many glacial lakes are dammed on one side by the ice itself, and the water may extend into crevasses and irregular-shaped holes in the wasting glacier. Such lakes are generally temporary, and fluctuate in level as movement or melting of the ice diverts their outlets. Sediments accumulate in them.

Because of fluctuations in lake level and movements of the glacier, these lake deposits are generally interlayered with stream deposits and with till. Upon melting of the glacier, patches of these sediments that filled former crevasses and holes in the ice are left in terraces and in isolated buttes called *kames* and *kame terraces*.

MODIFICATION OF TOPOGRAPHY BY GLACIERS

It is apparent that glaciers differ greatly from rivers in their mechanism of flow, and in their erosional and depositional features. The topography shaped by them contrasts sharply with the topography shaped by running water. Such differences are conspicuous in areas recently uncovered by glacial recession.

The polished and grooved bedrock floor is a feature not found in normal stream channels. Glaciated areas also contain swarms of lakes, ponds, and marshes that fill either shallow rock basins scooped out by the ice, or depressions dammed by moraines and outwash. The streams that connect these abundant lakes are totally ungraded. An immature stream pattern with numerous waterfalls and lakes is typical of recently glaciated areas.

The upstream sides of peaks, hills, or small rock knobs overriden by ice are commonly rounded, polished, and grooved; the downstream sides are irregularly jagged. It is inferred that the glacial rasp has sanded off the upstream side, but rock has been removed from the lee side, chiefly by quarrying of joint blocks caught by angular boulders dragged over them. Pressure relations may also be important. Pressure on the upstream face of the knob causes melting, and the water freezes in the low-pressure area, causing plucking on the lee side of the knob.

The usual mountain stream valley has a V-shaped cross profile but in valleys from which glaciers have melted the lower part of the profile is characteristically U-shaped, though remnants of a former V-shape may be preserved on surfaces higher than the top of the vanished glacier.

It is inferred that the spurs projecting between tributaries along the walls of a river canyon have been quarried away by the ice. A glacier cannot negotiate the sharp curves around spurs as easily as water; because of its bulk and brittleness it overrides the spurs and planes them into triangular facets forming a straight line of cliffs. At the same time, the floor of the valley is deepened by the glacial rasp. The widening, deepening, and straightening of former stream canyons by ice transforms the original V-profile to a U.

The long profile of a valley formerly occupied by a glacier is irregular and is commonly interrupted by abrupt steps above and below which the smooth U-shaped valley may continue with flat gradients. Several such "cyclopean steps" may appear in a single glaciated valley, alternating with polished rock floors and lake-filled basins. Valley deepening by glaciers is not controlled by a smooth grade profile related to sea level as in streams, but by the thickness of the ice, the speed and duration of its flow, and the structure of the rock in its bed. In restricted parts of its course, it is not unusual for a valley glacier to deepen its trough far below sea level.

Unlike stream valleys, glaciated valleys commonly head in cirques. Abandoned cirques are conspicuous features in recently deglaciated mountains. These gigantic semicircular basins may be so closely spaced that the divides between them are reduced to knife-edged comblike ridges, or to triangular mountain "horns".

Stream valleys and glaciated valleys also differ in the relation of the tributaries to the main stream. The surface of a tributary glacier may accord with that of the main glacier—though, because of the rigidity of a thin ice mass, this is not always so—but the floor beneath the glacier and tributary is cut to depths depending on the volume and speed of the respective ice streams and the kind of rocks over which they move.

The thin tributary glacier, unless favored by less-resistant bedrock, cannot keep pace with the scouring and plucking by the larger glacier. As a result, when the ice retreats, the tributary valleys are common left "hanging" high above the floor of the main valley, and their streams descend into the main valley in high waterfalls or steep rapids.

Now that it is obvious glaciers leave clear marks of their former presence upon a landscape, let us see if there is evidence of widespread former glaciation in areas that today have a temperate or even a tropical climate.

Former Periods of Glaciation

The Iowa farmer, sweating under an August harvest sun, may doubt that "the present is the key to the past" if he is told that the soil he tills has come from deposits left by an ice sheet that once covered most of northern North America in the Pleistocene epoch.

Similarly, the dark-skinned native of the Talchirs in India, who pauses to rest from the steaming tropical heat on a polished and striated ledge of rock, will doubtless consider the idea fantastic, though alluring, if told that his perch is the floor of an ancient glacier that spread widely over India in Late Paleozoic time.

Conclusions such as these so tax the imagination that, in spite of the clearcut evidence, they were not accepted even by geologists until every possible alternative—from the Biblical flood to some mysterious kind of volcanic activity—had been tested and found inadequate to explain the facts recorded in the rocks and in the drainage patterns upon them. Yet, evidence of the presence of former ice sheets can be readily gathered by any interested observer.

THE PLEISTOCENE GLACIATIONS

In most of Scandinavia, southern Canada, Labrador, and parts of northern United States, the normal soil profile formed by weathering is either poorly developed or missing. Instead, rounded hilltops expose smoothly polished rock ledges like those beneath present glaciers. In places, the polish and smaller striations have been destroyed by weathering, but where protected by even a thin veneer of till, outwash, or peat, they may be as fresh and clear as beneath a modern ice sheet.

Boulders—many of huge size—are scattered erratically over the polished surface. Many boulders are of rocks entirely different from the local bedrock. In the fields of the Iowa farmer, boulders of gneiss and granite rest on flat-lying limestones and shales.

The headwaters of streams in the area expose no such bedrock, so these boulders could not have been brought in by floods. In parts of Iowa, chunks of copper are occasionally found like the copper ore mined from rock ledges oil the Keeweenaw peninsula in Michigan or Isle Royale in Lake Superior, but not found elsewhere. Boulders of an unusual variety of granite called "rapikivi," known in outcrops only north of the Gulf of Finland, are scattered widely across Estonia and even far into Poland.

The basalt plateau of eastern Washington is strewn with huge granitic boulders of kinds found only many miles to the north. To reach their present position they must have been transported directly across the Columbia River canyon, which is here 1,500 to 2,000 feet deep. A large nickel deposit in northern Finland was actually discovered by tracing its strewn ore fragments northward to their source near Petsamo.

Development of the Glacial Theory

The implications of such relations seem obvious to us now. Nevertheless, in the period from 1821-1835, when two European geologists, Venetz and Charpentier, showed that boulders of rocks characteristic of the central Alps were widely scattered across the broad Swiss plain to the north, and correctly inferred a former extension of the present Swiss glaciers, their ideas met little but skepticism.

In 1836, however, Charpentier induced one of these skeptics, a young Swiss geologist named J. L. R. Agassiz, to accompany him on a visit to the active glaciers and, lower in the valley, the huge abandoned moraines of the Rhone glacier. Convinced on this expedition that the evidence for former more widespread glaciation was even stronger than Charpentier had claimed, Agassiz set to work on the problem.

He promptly saw and correctly interpreted the relation of the transported blocks to the polishing and grooving of the bedrock, and showed that these smooth striated forms could not possibly have been produced by water. Agassiz's work, too, met with general disbelief. As his biographer relates:

Men shut their eyes to the meaning of the unquestionable fact that... the former track of the glaciers could be followed, mile after mile, by the rocks they had scored and the blocks they had dropped.

Agassiz, however, was not easily discouraged. In 1840, he visited much of Scotland, the north of England, and parts of Ireland. There, he demonstrated glacial phenomena identical with those on the Swiss plain, and he announced that not only had glaciers once existed in Britain, but that they formerly covered most of the country. At first this conclusion raised a furore of objections, but it also set geologists to observing and more carefully analyzing the evidence. Faced with the facts so easily gathered over much of the British countryside, it was not long until Agassiz's views prevailed.

Agassiz emigrated to America and began the studies of glaciation in New England since followed so fruitfully by many investigators. Studies since Agassiz's time have supplied a wealth of information. Modern maps portray in detail the erosional forms and deposits of the former ice sheets.

Many abandoned moraines have been traced in the field, lobe by lobe, and plotted on maps until their distribution in North America and Europe is well known. It is not difficult to map the moraines. Although locally absent or poorly developed, most are conspicuous, easily identified, hummocky ridges of till that are readily traced for miles. Behind them, like signposts pointing the way, lie striated rock floors, locally strewn with till and foreign boulders.

Here, too, are innumerable lakes and marshes where the vanished glacier gouged its floor or blocked older drainage by debris it left on melting. In front of the moraines are coalescing fans of outwash and, still farther beyond, sheets of loess, or terraces of silt. All assist in locating the position of the edge of the vanished glacier, and all testify to its former existence. The landscapes of New York, New England, the Great Lakes states, and the Northwest are nearly everywhere stamped with these glacial imprints.

In mountainous areas, such as Yosemite National Park, the Cascade Mountains, or the Rockies, former glaciers are recorded by the numerous U-shaped valleys; the fresh moraines; the abundant cirques; the superb waterfalls pouring from "hanging valleys"; the clear mountain lakes nestled in rock-scoured basins; and the terraces of outwash gravel, sand, and silt extending downstream from the moraines.

Advance and Recession of Pleistocene Ice Sheets

Careful study of Pleistocene glacial deposits shows that they are not the result of a single advance and retreat of the ice. Instead, they record at least four stages of ice advance separated by long periods during which the glaciers withdrew to greater altitudes or latitudes than they occupy today, or even disappeared.

At many places in North America and Europe, roadcuts or natural exposures reveal two tills, one on top of the other. The upper layer of till may contain boulders of almost fresh granite and gneiss, some with polished facets and striae. In the lower till, however, beneath a layer of thoroughly weathered clayey soil, the outlines of similar boulders can also be made out, but the rocks have weathered to the consistency of cheese—the feldspars are thoroughly rotted to clay, and the ferromagnesian minerals have decomposed to limonite and other materials. Only chemically resistant rocks like quartzites are found intact in this lower till.

It is apparent that the older till had undergone long weathering and that typical A and B soil horizons had developed on it before the younger till covered it. The B-horizon of the weathered till is usually rich in clay that cements it into a tough sticky mass. Such tough clayey subsoils, whether of glacial origin or not, are popularly called *gumbo. Gumbotil* is the technical term used to designate deeply weathered clayey tills.

Moraines with mantles of gumbotil, unlike younger unweathered moraines, have generally been considerably eroded since they were formed.

Careful study of the degree of weathering of different glacial deposits and mapping of their mutual relations have permitted the discrimination of four main stages of ice advance. Between the ice advances the climate appears to have been mild and warm, by the fossils found in stream and marsh deposits between the tills.

DATING PLEISTOCENE DEPOSITS BY VARVED LAKE SEDIMENTS

Lake deposits are abundant in glaciated regions. Deltas grow rapidly into glacial lakes where torrents of turbid meltwater pour into them. The finest silt and mud settles slowly and is spread throughout the lake. Many glacial lakes are deep green, owing to the dispersion of light by the abundant particles of suspended clay.

In the winter, the lakes freeze over, and the small tributary streams may freeze solid. During this quiet period, the suspended clay particles beneath the ice settle slowly to the bottom. By spring, most glacial lakes have nearly cleansed themselves. In summer, however, the surface ice melts and the streams become gushing torrents of debris-charged water which sweep coarse as well as fine material out into deep water.

The coarse material settles promptly to the bottom, covering the fine winter deposit with a thin layer of coarse silt. Then, with the progress of the season, the silt is buried by successively finer deposits that gradually grade upward into the fine dark-colored mud of the next winter. In the summer, microscopic organisms flourish in the warm surface layers. They extract calcareous matter from the water. The carbonaceous bodies of the organisms sink only slowly on death. Thus, the summer layer is richer in light-colored silt, the winter layer in black organic matter. Such cycles have been observed to form in many existing lakes.

PRE-PLEISTOCENE PERIODS OF GLACIATION

Many ancient sedimentary formations are composed of unsorted debris showing all the characteristics of till except that they are tightly cemented. They contain striated and faceted stones and are associated with varved shales and slates or with sandstones and conglomerates showing the typical features of outwash deposits. These features, and their local occurrence above polished and grooved rock floors, leave no doubt that they were formed during an ancient glacial epoch. Such cemented glacial tills have been called *tillites.*

Although small bodies of tillite have been found in rocks of many different ages, two periods in the earth's history before the Pleistocene are characterized by particularly widespread glacial deposits. During the Late Paleozoic, ice sheets spread widely over India, South Africa, Argentina and southern Brazil, southern Australia, the Boston Basin in the eastern United States, and many other localities.

In India, Australia, and South Africa, grooved rock floors covered by the hard tillite may be seen at hundreds of localities. Their preservation today is due to the fact that they were buried under a succession of younger sediments. Deep burial protected them from weathering and erosion, and the unconsolidated glacial deposits were slowly cemented into rock. Recent uplift and erosion of the overlying deposits has now exposed them to view.

In the Late pre-Cambrian also, glacial climates appear to have been widespread. Ancient tillites of this age have been found on every continent except South America.

CAUSES OF GLACIAL CLIMATES

Geologists and climatologists have tried for more than a century to explain the recurrence of glaciation on a continental scale. Theory after theory has been suggested, but all explain too little or too much. None can be considered satisfactory, at least in its present form; yet they have an interest that justifies at least brief mention.

Facts to Be Explained

- Continental glaciers (in Greenland and Antarctica) occupy about 10 per cent of the land surface today. They formerly covered about three times as much area during four different epochs of Pleistocene time.
- Climatic zones during the Pleistocene were about parallel to their present positions but were displaced to the south during the glacial maxima and to the north during interglacial epochs. In low latitudes high rainfall was contemporaneous with glacial climates at higher latitudes.
- Estimates of the duration of the several glacial and interglacial stages of Pleistocene time do not suggest periodic recurrence; the climatic fluctuations seem to have been irregular.
- Evidence of continental glaciation on scales approaching that of the Pleistocene is also found in Late Paleozoic rocks (about 200 million years old) and in pre-Cambrian rocks (at least 500 million years old), but not in strata of intermediate age.

Suggested Explanations

The many theories that have been offered may be grouped under the following general categories: geographic, atmospheric, and oceanic; geophysical; astronomical; and cosmic.

The geographic theories attempt to account for glaciation by climatic changes that would be expected from changes in the areas of the continents with respect to the oceans, changes in oceanic and atmospheric circulation due to changes in the shape of ocean basins, mountain uplifts, and so forth. These causes are inadequate because there is no evidence whatever of any notable shift in the distribution of landmasses or of violent fluctuations in their altitude since Pliocene time, yet we have conclusive evidence of four great glacial advances and of milder-than-present climates in some, at least, of the interglacial times.

It has been suggested that changes in the amount of carbon dioxide and of volcanic dust in the atmosphere would bring about climatic variation. Carbon dioxide absorbs some of the heat radiated to space from the earth's surface. If its quantity were greater, more heat might be "blanketed in" and the temperature would rise, whereas, if it were less, the climate would become colder.

However, this mechanism is quantitatively inadequate, especially because the effect of simultaneous change in amount of water vapour would practically compensate for the variations in carbon dioxide. Volcanic dust undoubtedly screens out some of the sun's radiation, but there is no sign that volcanoes were more active during the stages of ice advance than during interglacial times. Changes in salinity of sea water (and consequently modifications of the oceanic currents and their climatic influences) are likely results from glaciation but they can hardly have brought it about.

Geophysical theories that attribute glaciation to shifts in the positions of the continents are obviously not applicable to the Pleistocene climatic fluctuations, nor is there any strong evidence that these could ever have sufficed for earlier glaciations. Although it is conceivable that the amount of heat given off from the interior of the earth may have fluctuated from time to time, measured fluctuations are so small compared to the heat received from the sun (less than 1/10,000) that they seem negligible as a climatic factor.

The astronomical theories take account of the periodic variations of the earth's motion with respect to the sun. There are several such variations and there can be no doubt that they would have some climatic effect. Each affects the distance of a particular point on the earth's surface from the sun and hence the amount of solar radiation it receives. They could not have alone produced the glacial periods, however, because their effects on northern and southern hemispheres, though not always exactly opposed, would not be identical, either. Yet, we now have recession of glaciers taking place simultaneously in both hemispheres and it is reasonable to think that earlier fluctuations were also simultaneous.

Furthermore, since all these astronomic motions have been going on through geologic time, we should have had continental glaciation every few hundred thousand years. Instead, we must go back to the Permian to find anything comparable to the Pleistocene glaciation in the geologic record. These geologic arguments seem conclusive, but several meteorologists contend, in addition, that the temperature changes produced by this mechanism would be entirely too small to produce glaciation even in one hemisphere.

The cosmic theories are of two kinds:

- That the solar system from time to time encounters cosmic dust clouds
- That the sun varies in the amount of heat it radiates.

The cosmic dust clouds may either screen out the sun's radiation (if the component particles are fine enough) or blanket the earth's radiation to outer space (if the particles are somewhat larger). This is, of course, a very flexible hypothesis, as there is no way of testing it; with one assumption we can get a colder climate, with another a warmer. Accordingly, this idea is not seriously considered by most geologists, even though it may be the true explanation, for the scientific approach is to consider seriously only those hypotheses that are susceptible of testing in some fashion, however indirectly. The second suggestion, that the sun's radiation fluctuates somewhat over long periods of time, is on a better footing.

The sun's radiation does, in fact, fluctuate, and with it the amount of heat received by the earth. Variable cloudiness and other atmospheric factors make accurate measurements difficult, but short-term variations of as much as 3 per cent in the heat output seem well established.

The American astronomer, Abbot, a leader in this research, has declared that larger solar changes may well have taken place in the past, even though it is not yet possible to prove it. Several climatologists have tried to deduce the consequences of such moderately large changes, if they do, in fact, occur. We have not space to review all the hypotheses that have been suggested; perhaps the one that has won most support from meteorologists is due to the British meteorologist, Sir George Simpson.

Simpson (1934) reasoned that if solar radiation increased, the air temperature would rise all over the earth, winds would increase and so would cloudiness and precipitation. The increased precipitation would lead to greater accumulation of snow and the increased cloudiness would lessen summer melting.

The ice caps in both hemispheres would expand. However, the rising temperature would eventually prevent further advance of ice and ultimately cause melting. At the high point on the radiation curve, the world climate would be milder and wetter (more "oceanic") than it is today. When radiation again decreased, the reverse sequence would be followed: appearance and advance of the ice; then retreat because of lower precipitation; and, finally, when radiation fell to its present value we would be back to present conditions, with polar accumulations of relatively small size.

Thus, the somewhat paradoxical condition would arise, that a rise in the general temperature of the air would, by increasing precipitation and cloudiness, lead to continental glaciation of much of the present temperate zone. On this theory, one rise and fall of radiation would produce two glaciations in high latitudes, separated by a warm and wet interglacial time.

The climate following the second glacial would be cold and dry like the present. In low latitudes there would be a single epoch of increased rains that would endure through both high-latitude glacials and the intervening warm, wet interglacial.

The four Pleistocene glaciations recognized in Europe and North America would, on this hypothesis, require two cycles of increased solar radiation, separated by a time during which the climate may have been much as it is today. Simpson computed that, during the maximum of solar radiation, the mean temperature was 5° to 10° C. warmer than at present and that the average cloudiness was increased from the present 54 per cent to between 70 and 80 per cent. Simpson computed that these changes are reasonably to be expected if the sun's radiation fluctuated about 20 per cent above its present value.

This hypothesis is attractive because it would account for the increased precipitation that would be necessary to feed ice sheets. The present line of mean annual temperature at the freezing point lies far closer to the equator than the limit of present ice sheets, and this clearly indicates that low temperature alone is not enough to produce glaciation but that increased precipitation is needed.

With the present configuration of land and sea, which is not greatly different from that during the Pleistocene, increased precipitation can only be brought about by increased evaporation from the oceans. The theory also accounts for the warm, moist climate suggested by plant fossils from some interglacial deposits. On the other hand, there is no general evidence that the last interglacial period was generally warmer and moister than the second, as it should have been on this hypothesis.

Nor is there general agreement that the increased rainfall in lower latitudes lasted throughout both the last two glacial epochs and the intervening interglacial epoch. Finally, inasmuch as most astronomical processes are cyclic, if the variation of the sun's radiation is also cyclic, we should have had numerous other glacial episodes throughout geologic history. We must conclude that, even if this theory is satisfactory in some respects, an auxiliary hypothesis is needed to account for the long nonglacial periods of the geologic record.

We thus see that no satisfactory hypothesis has yet been advanced to explain the widespread glaciation of the continents during the

Pleistocene. It is impossible to say whether or not the present is simply another interglacial epoch. We cannot tell whether glaciers will again encroach on the sites of Chicago, Copenhagen, and Warsaw within a few thousand years, or whether the climate will remain about the same as it is today. Much more work will have to be done to win an answer to this fascinating problem.

THE EFFECT OF GLACIAL LOADS ON THE EARTH'S CRUST

There is strong evidence from deflection of the plumb line and from gravity measurements that large segments of the earth's crust are essentially in isostatic balance. Glaciers of continental dimensions also afford clear geologic evidence in support of this conclusion.

The fact that former glaciers flowed over and eroded the ground on which they rested is proof that their surfaces sloped down from their feeding grounds to their termini. Wherever the glaciers spilled around mountainous topography, it is possible to get some measure of their thickness. In New England, for example, the highest peaks were unglaciated, and a measure of more than 4,000 feet has been made for the Pleistocene ice in that region. In southern Canada the ice would certainly have been considerably thicker.

Although the density of ice is only a little more than one-third that of ordinary rock, a load of 3,000 feet of ice over an area as large as the Great Lakes and southern Canada would be the equivalent of a load of more than 1,000 feet of rock placed on the area, and should bend the crust down. Though the isostatic response by subcrustal flow might take considerable time, owing to the high viscosity of the subcrustal material, the lowering should be considerable and measurable, if isostasy really is general.

Fortunately, the shorelines of ancient glacial lakes give us a means of testing the idea. They were level surfaces when formed. After the ice load melted away the rise of the crust in response to unloading should tilt the shorelines, causing them to rise to the north. This is exactly what is found, not only in the Great Lakes region, but in the Baltic, in Labrador, and even in the basin of Lake Bonneville, where the load was only 1,000 feet of water covering a much smaller area than that of the great continental glaciers. Suffice it to say here that the testing of the isostatic principle by glaciers is in full agreement with the geodetic evidence from which the theory was first announced.

Chapter 9

Formation of Thunderstorms

Thunderstorms occur most frequently in the warm, moist, and unstable air masses within the tropics, where (in the rainy season) they are a daily afternoon occurrence, and sometimes so regular that they serve as a convenient time indicator for human affairs. In arctic regions, where the air is cold and comparatively dry, thunderstorms may not occur oftener than once in several years. Temperate regions experience their share of storms in the late spring and summer. No part of the United States is free from thunderstorm activity, but different regions vary widely in thunderstorm frequency.

In winter the centre of electrical activity is over the lower Mississippi Valley; in summer there are two far more active centers, one over Florida and the eastern Gulf coast, the other over northern New Mexico. But these centers are only relative, of course, for many parts of the nation see nearly as many storms. And in some regions, such as the Central, North Central, and East Central States, thunderstorms are likely to be mostly of the frontal type, and hence more severe and destructive.

Thunderstorms are classed, according to their apparent severity, as 'mild,' involving few if any ground-strikes and relatively light winds; 'moderate,' involving fairly frequent groundstrikes, moderate to heavy rain, and moderate to strong winds; and 'severe,' attended by nearly incessant lightning, torrential rain, and severe squalls. Temperate-zone thunderstorms are also classed, according to the mechanism of their formation, as 'airmass' and 'frontal.' The older names for these same classes of storms, before air-mass-and-front analysis came into wide use, were 'convective' and 'cyclonic.'

Air-mass thunderstorms occur entirely within one air mass, and without aid from any frontal action between air masses. They arise where the characteristics of the air mass, aided by wind and topography, favour them. They are most frequent in the United States in the warm, moist, and conditionally unstable air masses of tropical maritime origin.

But moist polar air masses advancing southward rapidly, and hence remaining relatively cold in their upper levels, can occasionally be unstable enough to produce scattered air-mass thunderstorms.

The atmospheric instability that produces air-mass thunderstorms and keeps them active for hours over land areas in daytime is usually 'triggered off' by intense solar heating at the surface, which steepens the vertical temperature gradient; or perhaps by the importation of cold air aloft; or by the rising currents produced by upward deflection of horizontal winds (together with perpendicular solar heating) on the slopes of hills or mountains. Thus air-mass storms are frequent along the Gulf coast, where the sea air strikes the hot and slightly rising land; and thunderstorms of any type are more frequent over mountains than over flat country or at sea.

Frontal thunderstorms may occur along a warm front, where even gradual lifting on the warm-front slope is sufficient to trigger off the instability of very warm and very moist tropical air. They may also occur along the wave-like oscillations of a more or less stationary front that is, for the moment, alternately warm and cold. They may even arise, rather mysteriously for unaccustomed forecasters, out of upper fronts that give little or no other indication of their motivating presence.

But the great majority of frontal thunderstorms occur either directly in front of, or actually ranged along, a vigorous cold front marked on the weather map by large temperature contrasts, a deep pressure trough, and strong shifting winds (that is to say, by a fairly strong pressure field having isobars sharply bent into V-shape where they cross the front). This whole formation is, of course, nothing more or less than a line squall.

Thunderstorms of all types are most frequent, on land, during the middle of the afternoon — that is, during the hours when solar heating at the surface reaches its maximum and the lapse rate (vertical temperature gradient) is consequently steepest. As the average life of a thunderstorm is six or seven hours, however, thunderstorms formed in the afternoon often persist into the early evening.

Over the ocean, where the equable water warms up by only a degree or so during the day, and cools off by even less at night, the thunderstorm maximum comes during the late night and early morning hours, when radiational or airmass-transport cooling aloft is greatest, and, again, when the total lapse rate is steepest.

As the mechanics of frontal thunderstorms depend not so much on surface heating or upper cooling as on the violent upthrusting of warm and moist air masses by an undercutting wedge of cold air, frontal storms may occur at any time of the day or night. But even the frontal storms are most likely, and most severe, during the middle of the afternoon on land or during the early morning hours at sea. Plenty of active summer cold fronts, for example, do nothing much over land during the morning hours, yet develop into roaring line squalls by mid-afternoon.

FORMATION OF A THUNDERSTORM

What are the aerological conditions that favour the formation of a thunderstorm? First, there is the all-important factor of atmospheric instability, evidenced by a steep or steepening lapse rate (-3° F/1000 ft or steeper, or about -6° C/km or steeper). This steep temperature gradient is perhaps being increased by air-mass transfer below or aloft — lower warm winds from the south and southeast and/or upper cold winds from the north and northwest — and above all, by excessive solar heating at the surface.

In addition to the steep temperature gradient, high relative humidity at all levels (meaning that clouds will be formed by very little rise of air) also favors the formation of a storm. Finally the 'ice level,' or altitude of freezing temperature (+32° F. or 0° C.), should be located at considerable elevation, as it always is over the tropics.

These upper-air characteristics, of course, can be fully known in advance only by means of airplane or balloon soundings, though they can sometimes be sensed by the rapid (and occasionally prodigious) vertical growth — the spectacular upward mushrooming — of detached cumulus clouds during the late morning.

There are also some surface thunderstorm portents not without value — high temperatures, say 80, 90, or 100° F. by day and 70° or 80° by night; high surface specific humidity as indicated by high dewpoint, say more than 6o° or 70°; and steadily decreasing surface pressure during the late morning and early after. noon.

Another useful early-afternoon indication is increasing 'crash' static from a radio set — actually radio waves from lightning flashes beyond the horizon. Finally there are the definite evidences of frontal activity and advance towards one's position to be obtained from some sort of weather map — even a simple regional map that anyone can draw from airways radio weather reports in a few minutes' time.

One or more of the indications just listed can of course be present without resulting in thunderstorms, and thunderstorms can occur when one or more of them are absent; but in general the presence of most of them makes thunderstorms likely, and the agreement of all of them makes thunderstorms certain.

Given these favorable conditions, how does an air mass

thunderstorm actually form and grow? The beginning of a thunderstorm is similar to the formation of an ordinary cumulus cloud or an ordinary shower. But in the case of the thunderstorm the air is more moist and more unstable, and this moist instability extends to greater heights. Hence the updrafts are more violent, and the cloudbuilding is more rapid and extensive.

As in ordinary cumulus-building, the rising currents of a thunderstorm cause the warm, moist air to cool by expansion to its dewpoint, forming minute liquid droplets (on commonly present condensation nuclei), perhaps 1/2500 inch in diameter, which droplets appear as the white or gray opacity of the cloud; and this condensation of water vapour, in a vicious cycle, releases latent heat which helps to drive and intensify the rising currents, which condense further moisture, and so on. The result is that under favorable conditions the thunderstorm grows fast and far — perhaps up to altitudes of six or seven miles.

Just as in the formation of all rain, so, while the minute cloud droplets of a thundercloud are carried up in the rising currents, there is some jostling between them, which jostling tends to coalesce many of them into much larger drops. These larger drops, if the updraft is not too strong, will begin to fall through the cloud as rain. If the upsurging cloud droplets reach a temperature as low as +5° F. (-15° C.), which they soon do in a

by high dewpoint, say more than 6o° or 70°; and steadily decreasing surface pressure during the late morning and early after. noon. Another useful early-afternoon indication is increasing 'crash' static from a radio set — actually radio waves from lightning flashes beyond the horizon. Finally there are the definite evidences of frontal activity and advance towards one's position to be obtained from some sort of weather map — even a simple regional map that anyone can draw from airways radio weather reports in a few minutes' time.

One or more of the indications just listed can of course be present without resulting in thunderstorms, and thunderstorms can occur when one or more of them are absent; but in general the presence of most of them makes thunderstorms likely, and the agreement of all of them makes thunderstorms certain.

Given these favorable conditions, how does an air mass thunderstorm actually form and grow? The beginning of a thunderstorm is similar to the formation of an ordinary cumulus cloud or an ordinary shower. But in the case of the thunderstorm the air is more moist and more unstable, and this moist instability extends to greater heights. Hence the updrafts are more violent, and the cloudbuilding is more rapid and extensive. As in ordinary cumulus-building, the rising currents of a thunderstorm cause the warm, moist air to cool by expansion to its dewpoint, forming minute liquid droplets (on commonly present condensation nuclei), perhaps 1/2500 inch in diameter, which droplets appear as the white or gray opacity of the cloud; and this condensation of water vapour, in a vicious cycle, releases latent heat which helps to drive and intensify the rising currents, which condense further moisture, and so on. The result is that under favorable conditions the thunderstorm grows fast and far — perhaps up to altitudes of six or seven miles.

Just as in the formation of all rain, so, while the minute cloud droplets of a thundercloud are carried up in the rising currents, there is some jostling between them, which jostling tends to coalesce many of them into much larger drops. These larger drops, if the updraft is not too strong, will begin to fall through the cloud as rain. If the upsurging cloud droplets reach a temperature as low as +5° F. (-15° C.), which they soon do in a towering cloud (at a height of four or five miles in temperatezone summer), some of them are jostled into ice particles; and ice particles are perhaps also subliming directly out of water vapour (onto sublimation nuclei already present at these cold heights), if the temperature and humidity conditions are favorable.

In a fully developed thunderstorm the ice-particle cloud is often visible as a cirrus-like, anvil-shaped top or cap that blows out somewhat ahead of the main cumulus turret on the upper winds. The myriads of ice particles, falling slowly and growing into larger crystals, eventually produce rain and other forms of precipitation.

In this fashion the production of rain continues until large drops perhaps a quarter of an inch in diameter are falling through the cloud; and the rising currents build up until, in the forward and most active part of the storm, they exceed twenty miles an hour. This updraft will just support a three-eighths-inch raindrop; and no raindrop larger than three-eighths inch can exist in nature, for it would be instantly blown into smaller drops by the wind of its fall.

Thus no rain can fall through the most active updrafts of the thundercloud, and the active front of the storm is marked by an abrupt stoppage of large-drop rain falling from above, by the deflection of this rain upward towards the rear of the storm, and by the breaking up of large drops into smaller drops.

From the breaking up of these water drops, which can, of course, recombine above and descend to be broken again, come the electric charges that produce the lightning. At the instant of spattering, the to-be-broken drops are charged with positive electricity, and the air surrounding them is negatively charged. This negatively charged air is carried away into upper and rearward portions of the thundercloud, but the positively charged raindrops, breaking and recombining, tend to accumulate in the most active (forward and lower) portion of the storm, which hence becomes a region of high positive electrical charge.

From this positively charged region, usually just above and behind the roll cloud or 'squall head,' frequent heavy thunderbolts strike downward and forward directly into the ground (whose charge is neutral or somewhat negative), while other longer but less intense flashes strike upward and backward to lose themselves in the upper immensities of negatively charged cloud.

Towards the rear of the storm, however, where the earth is positive with respect to the negative clouds above, less intense bolts strike upward from the ground into the clouds.

Lightning may be only one feature of a great storm perhaps marked also by cloudbursts, ice showers, and destructive blasts of wind, but it is spectacular enough, and under some conditions hazardous enough, to warrant examination in some detail before considering the other aspects of the storm.

THOR'S HAMMER

The potentials set up by the cloud-formed electric charges of a thunderstorm are enormous — in the neighborhood of one hundred million volts — and they discharge through dry air, ordinarily considered one of the best insulators, to usual distances of half a mile, or even occasionally to a mile or more. The lightning bolt itself, though similar in many ways to a small spark in the laboratory, yet shows additional and peculiar characteristics introduced by its enormous voltage and the great length and diversity of its path.

There is nothing oscillatory or alternating about the lightning bolt — it is one or more surges of current in the same direction. Moreover the bolt does not jump instantly from a positive cloud, say, to the negative earth — it appears to grow or burrow through the air in a succession of electronic darts, building its own ionized (or electrically conducting) path as it goes.

Ordinarily this growth is frightfully rapid, and takes place in perhaps one millionth of a second. But occasionally it is slow enough to be seen by the naked eye in the form of 'rocket lightning,' in which a growing and perhaps upward-directed bolt burrows visibly at no greater speed than an ordinary rocket. Some of the veined lightning in a detached thunderhead, viewed from several miles away on a clear night, can often be seen to grow or burrow in this fashion progressively along the bulges of cloud. Usually lightning strikes from thundercloud to ground, from ground to cloud — or, even more commonly, between clouds. But on rare occasions it flickers or glows out of a clear sky, as charged masses of air pass near each other or the ground.

The amperage or rate of electric current flow in a thunderbolt is likewise enormous, compared with most man-made currents. Its value ranges all the way from ten thousand amperes in weaker flashes up to perhaps two hundred thousand amperes in larger bolts. One hears of people being struck by lightning and surviving the ordeal, but these fortunates are probably grazed by side flashes or forked branches while the main bolt goes elsewhere.

Few if any animals, human or otherwise, have ever been hit by a heavy main bolt and lived to tell the tale. For like its voltage and amperage, lightning power is unbelievably large — perhaps ten million million kilowatts or thirteen trillion horsepower! But as an average single flash lasts only three or four millionths of a second, the total amount of electricity (or electrical energy) involved is not very large — perhaps three thousand kilowatthours or so, obtainable commercially within more usable limits of current and voltage for seventy to a hundred dollars.

Lightning punishment at one point is by no means limited to a single bolt, however, and the old adage that 'lightning never strikes twice in the same place' is about as untrue a statement as anyone could conveniently concoct. Many lightning flashes are of the 'multiple' variety; the bolt not only strikes twice, but may strike repeatedly — five times, ten times, or even forty times in the course of a second or two. This multiple lightning usually appears to the eye as a single, flickering, long-lived bolt, and it is the kind most likely to start fires in any dry and combustible material along its path.

The brilliant flash that accompanies a lightning discharge, formerly thought to be caused by sudden heating of air, is now known to result from ionization of the air by the terrifically strong electric field along the path, causing the very atoms of dense air to vibrate at such high frequencies that they radiate visible light in the same way that rarified neon in a glass tube glows under moderate electrification.

The lightning flash is thus kindred (though once or twice removed) to the flickering aurora of polar skies, and to the glowing message of an advertising sign. Lightning is usually white in colour, combining the spectrum of nitrogen with that of oxygen. But it may appear bluish in contrast to yellowish artificial lights, and may appear yellowish if viewed through haze, fog and clouds. Occasionally a reddish or pinkish flash occurs, when air rich in water vapour is ionized so as to give the spectrum of hydrogen. Thunder takes many audible forms, but they all originate in one underlying cause — the sudden increase of air pressure along the path of the bolt (for any sound wave is only a sudden increase in air pressure) resulting from the heating, molecular dissociation, and ionization of air by the bolt.

Earlier vague assertions about 'air collapsing to fill up the hole hollowed by the bolt,' and such notions, are now in the discard. If lightning strikes close enough — within a few feet of us — the sound is simply one sharp and terrific 'bang!' like that of a large cannon fired in the same place; for the ear is too paralyzed by this first and nearest blast to notice the subtler and later gradations of reverberating thunder arriving from more distant portions of the bolt.

But if the bolt is farther away, and particularly if it is a long one, the thunder arriving successively (at eleven hundred feet per second) from its more and more distant portions has the sound of a long roll, which wavers considerably even as the characteristics of the bolt, and its direction with respect to the hearer, vary greatly along its path.

Under certain conditions of air stratification, and often in the presence of steep mountains and cliffs that also serve to reflect and intensify sound, the roll of the thunder echoes back and forth; these overlapping echoes, together with simultaneous sound reinforcements from equidistant portions of the path, may build up into great booming crescendos of noise. Thunder is seldom heard over horizontal distances of more than fifteen miles; and if the flash occurs at altitudes of more than two miles, it may not reach the ground at all in audible form.

Several kinds of lightning have been observed, of which by far the most common is 'streak' lightning — the ordinary bolt. A streak of lightning is not a zigzag, as commonly drawn, but sinuous in shape, like a river in broken country. The streak may be single, but more commonly, part of it splits off into smaller downward branches that may or may not reach the ground.

Occasionally a bolt splits into 'forked' lightning — two or more main branches that each strike home. Multiple streak lightning, drifted sideways by a strong wind, sometimes appears as 'ribbon' lightning. And if a single streak follows a very sinuous and varied path, it may present the illusion of glowing beads where it is either more intense, or else where part of the flash is seen end-on.

Another distinct variety, very rare but very dangerous, is 'ball' lightning. Scientists long doubted altogether the existence of this variety, but so many accounts of its appearance were authenticated that it has come to be believed in, and more or less explained. It appears in the form of a luminous ball, perhaps the size of a clenched fist, which either falls through the air, floats horizontally, or rolls along some solid object such as the ground.

Visible only for a second or two, it disappears either quietly, or with a terrific explosion. According to present theory, ball lightning is nothing more than a creeping corona discharge along a relatively intense portion of the electric field — or, one might say, a 'stalled thunderbolt,' similar in kind to rocket lightning but much slower. If the electric charge behind it dissipates, the ball fades away also; but if the original charge increases enough to break down the whole path with a large thunderbolt, as it well may — then look out!

Another variety of Nature's electric illumination is 'sheet' lightning. This is not to be confused with so-called "heat' lightning — the diffuse and silent reflection of distant electric flashes hidden either by clouds or by the horizon. True sheet lightning occurs as a momentary glow extending through fairly large masses of air in the cloud region (or troposphere), and is nearly always white in colour. It is probably an ionization effect similar in principle to the aurora, though the sheet-lightning glow occurs on a much smaller scale than the aurora, and at much lower levels in the atmospheric ocean.

Another and somewhat similar variety of thunderstorm light is ' Saint Elmo's fire,' a bluish, brush-like discharge that sometimes appears on rooftrees, tall steeples and chimneys, the mastheads of a ship, or any other high, sharp-pointed object during a thunderstorm. It is merely the well-known corona discharge, reproducible at will in the laboratory or on a high-voltage transmission line.

Occurring in nature, it indicates intensification of the electric field, and may indicate the imminence of a thunderbolt. Such a corona discharge, of course, presents a very real fire hazard — particularly in the case of a hydrogen-filled balloon or airship. On May 6, 1937, during a great line squall over New Jersey, the German airship *Hindenburg* was valving explosive hydrogen in preparation for a landing at Lakehurst. Saint Elmo's fire ignited the escaping hydrogen-air mixture, and the great airship crashed in flames, killing thirty-six of its transatlantic passengers.

This incident marked the tragic end of hydrogen-filled airship activity; and as the United States Government wisely declined to release helium to a nation arming to the teeth for a treacherous and unprovoked attack on its neighbors, it marked the effective end of all German airship activity for some time.

LIGHTNING HAZARDS

Just as lightning bolts themselves exhibit almost endless varieties of form and intensity, so do their earthly effects vary greatly, not only with the bolts' severity, but even more with the things and substances which a ground-strike encounters on its path to earth.

Metallic conductors of sufficient size — say a bar of copper or other metal the size of a pencil — pass the bolt without harm to themselves or anything else. But if the con- ductor does not offer the shortest and easiest path to earth (from the viewpoint of the hop-along bolt itself), it may jump to a more tempting and destructive path; and even if the main bolt follows the conductor, side flashes may jump to neighboring conductors, be they metal tanks or pipes, pools of water, or human beings.

Small, hollow conductors may be crushed by the intense electrostatic and electromagnetic fields accompanying lightning passage. Irregularly shaped, loose, or insecure rods or wires are likely to be moved, bent, burned, or melted; and insufficiently large conductors, such as a radio antenna, are invariably melted or vaporized instantly by a direct stroke.

Non-conductors such as wood and stone fare much worse, at least during the 'dry' part of the storm before they are wetted over with a film of partly conducting water. Dry trees and poles are often split and splintered throughout their length by the explosive action of their moisture content suddenly converted into steam, and the pieces are blown forcibly outward in all directions. Dry or partly moist bricks and mortar may be violently exploded, and very rarely a small and flimsily built house is wrecked completely, as if by an earthquake.

Some of lightning's worst destructive effects are seen where large conducting masses are separated by non-conducting ones -such as a metal roof, ungrounded, on top of a wooden house. Here the conductor seems to collect and intensify the bolt, which wreaks havoc on the non-conducting structure beneath. No ordinary insulator such as a glass plate, of course, offers any effective resistance to a lightning bolt. If an insulating plate stands in the way and cannot easily be got around, the bolt simply blasts or burns a hole right through it.

The best possible form of lightning protection for any object, animate or inanimate, is a 'Faraday cage' — in complete form a grounded shell of metal not less than one twentieth inch thick (Number 16 A.W.G.), completely surrounding the object in question and several feet away from it on all sides. But as such a metal box would be very expensive and inconvenient, it can be approximated, with nearly as good practical results, by a skeletonized cage of interconnected conductors — each conductor being at least one fourth inch thick, if of copper; or one third inch thick, if of iron.

Of such general form are the lightning-rod systems now in use on thousands of isolated houses and barns in thunderstorm areas — systems that prove their usefulness by greatly reducing fires and other lightning hazards. In general, these protective rods should extend at least a foot or two above all the higher projecting parts of a building; they should extend downward as directly as possible to thoroughly good ground connections; and they should be bonded to each other and to any other large metallic masses in the building, which latter should be independently grounded.

Any high metallic conductor, such as a steel tower or a steel-frame city skyscraper, usually confers about it a 'cone of protection' having a radius equal to three or four times the height of the tower; occasionally, however, this cone of protection is violated by an erratic bolt.

But for all the unpleasantness of lightning in the open — the vivid flashes of uncomfortably close ground-strikes and the deafening crash of near-by thunder — the actual hazard is really very small. Out of all the millions of people exposed to lightning in the United States, only about five hundred are killed by it each year. Under the intense part of a very severe storm, however, and in various unwise locations beneath any storm, the risk may notably increase. The Bareau of Standards has issued some rules of wise personal conduct for those caught out in the open by storms, which are as follows:

- Do not go out-of-doors or remain out during thunderstorms unless it is necessary. Stay inside of a building where it is dry, preferably away from fireplaces, stoves, and other metal objects.
- If there is any choice of shelter, choose in the following order:
  - Large metal or metal-frame buildings.
  - Dwellings or other buildings which are protected against lightning.
  - Large unprotected buildings.
  - Small unprotected buildings.
  - If remaining out-of-doors is unavoidable, keep away from:
  - Small sheds and shelters if in an exposed location.
  - Isolated trees.
  - Wire fences.
  - Hilltops and wide open spaces.
- Seek shelter in a cave, a depression in the ground, a deep valley or canyon in dense woods, or a grove of trees.

Their cumulus clouds are merely prosaic banks of condensers, but their lightning is real and convincing, as anyone can testify who has seen one of their fivemillion-volt flashes strike downward in the dark through fifteen feet of air to shatter a three-inch fence post with the bang of a small cannon.

Safety of People in Ordinary Wood-Frame Houses

Much will depend upon the construction of the house and what is in it, and particularly the amount of metal in the walls, whether in the form of gas, water, or vent pipes, how the electric wiring is arranged — whether in rigid conduits or in BX, or knob-and-tube open wiring. Quite obviously, the only absolutely safe place to be is inside of a completely enclosed metal structure, which may or may not be grounded.

The degree of safety enjoyed by the occupants of an ordinary wood-frame house will depend upon the degree of approach to this completely shielded structure. The wave front of current discharge in a direct stroke is steep enough, and the magnitude great enough, so that in many cases fairly high potentials will be built up in conductors of thirty- to forty-foot lengths — so that side-flashes may occur to other metal structures which may be connected to the same ground but at some distance away.

This was particularly exemplified in a case where a man had a lightning rod system properly installed, as far as the lightning rods were concerned, and yet it flashed through about six inches of wood and air to a conduit located in the peak of the attic roof underneath the lightning rod system. Of course if these two had been tied together at this point, the fire which resulted would have been avoided.

So far as my experience goes, the greatest danger from lightning in the ordinary wood-frame house is from fire, and of course this is particularly so in the rural communities where fire-fighting apparatus is not readily available. In cities the damage done to buildings as a result of lightning is not as a rule serious, although cases are on record where explosions have occurred during lightning storms, but apparently these explosions were due to the existence of gas or other explosive materials enclosed within the house at the time it was struck.

Wood-Frame Houses Protected by Lightning Rods

Insurance records for the State of Iowa indicate a very substantial improvement, which is also substantiated by those from the Province of Ontario, as reported by the Fire Marshal of Ontario. These results have indicated that lightning protection is better than ninety per cent efficient, from the point of view of preventing fires in protected buildings.

Of course not all of the lightning protection systems were properly installed. Many were, no doubt, inadequately grounded, which accounts for the imperfect record to some degree. At the same time, we know today that lightning discharges do not always strike the highest object, and that now and then the cone of protection, which laboratory tests have demonstrated exist about a lightning rod, is violated. I have, for instance, a photograph showing lightning striking the top of the Empire State Building, in which the discharge failed to contact the very top of the building but struck about forty feet down. It is apparent that lightning discharges take a path which might well be called the best path, from the point of view of the discharge itself.

Safety of People in Closed car with steel Turret top

Following the principle of the Faraday cage, there is no reason why people inside of a car with a steel top should not be very safe from lightning discharges, provided that they are not in the process of getting in or getting out of the car, or do not have their heads or arms out of windows, or are not standing on the ground changing a tire, and assuming that the gas tank is not struck and punctured, which is relatively unlikely.

I believe the best advice to a man riding in an automobile during a lightning storm is to keep going and not to get out of the car, and the chances are that he will be perfectly safe. Of course, if power wires fall on the car, which has happened during violent storms, he is wise to remain in the car, and not attempt to get out, since this procedure has in the past resulted in serious accidents.

Safety of People in Closed Wooden Yacht

As a general thing, lightning is more likely to strike a wet mast on the water, if it is striking round about, than a similar object on land, because of the relative smoothness of the water and the relative height of the projecting mast. It is probable that suitable protection could be provided to the boat and its occupants by connecting a conductor following down the mast to the metal keel of the boat.

In case it is an engine-driven boat, care should be taken to be sure that all the metal objects within the boat which are not connected together are electrically bonded together, and connected to some outside metallic conductor, such as the keel, which furnishes good contact with the water.

I don't know of any definite rule which has been set up concerning the protection of boats, but it seems to me that the general principle of the Faraday cage, or its equivalent, is one which can be applied to boats, coupled with the idea of the lightning rod and its usual cone of protection, and means for discharging the current of the lightning discharge itself into the water, without having either the boat or its occupants become a part of the circuit.

It might be well at this point to inject an observation concerning people in swimming, to the effect that, in my estimation, the danger to the swimmer is not so much that he may be struck by lightning, as that a small current flowing through the water, even at considerable distance from the stroke itself, may be sufficient to render him unconscious, and he dies by drowning, rather than by electrical shock.

Safety of People in an Airplane

Once more the question is one of preventing the passenger or pilot from becoming a carrier of current of the lightning discharge by being in series with it. The only known way to prevent this is either to keep the lightning current from entering the plane by means of an all-metal plane, or its equivalent protection by some means of shunting the discharge which gets into the plane from the body of the occupants, and means also of preventing the lightning discharge from interfering with the mechanical and electrical operation of the plane.

The best method of doing this is probably the all-metal plane, which brings us to the Faraday cage, where if the joints in the plane are thoroughly bonded together, and the metal parts so connected that the measuring instruments and other electrical circuits of the plane are not likely to become a part of the lightning path, the possibility of damage is relatively small. The occupants of the plane ought to be perfectly safe provided they are entirely surrounded by metal having a reasonable current carrying capacity....

From the point of view of fire, the multiple stroke is undoubtedly far more serious than the single stroke, and I believe I am quite safe in stating, that the single stroke of large magnitude is probably responsible for the explosive effects frequently observed, whereas the multiple stroke, even though of many relatively small currents, may be responsible not only for explosive effects but for fires also, provided inflammable material is present. As many as forty strokes have been observed within one second of time, and data which we have been taking during the past two years seem to indicate by far the great majority of lightning strokes are multiple....

DOWNBLAST AND DELUGE

Earlier in this chapter we saw how a typical thunderstorm forms out of nothing into a towering Everest of threatening cloud, how the terrific electric potentials of the lightning arise out of a mere spattering of raindrops. But what about the lashing hail, the scouring squall winds, and the other thunderstorm accompaniments that often prove more dangerous and destructive than the lightning itself?

We had earlier in mind the picture of a towering cumulus cloud marked by fast-rising air currents and fast-falling raindrops of near-maximum size, whose spattering in the zone of strongest updrafts produced positive electric charges large enough to forge the thunderbolts.

As most of the warm, moist air contributing its latent heat to these rising currents is coming from in front of the storm, the most violent updrafts are usually in the front part of the main thundercloud; and they tend to slope somewhat towards the rear of the storm, carrying the rain rearward also. Just behind the main updrafts, where their speed has fallen off to less than twenty miles per hour, will be the region of heaviest rainfall.

Much of this heavy rain, coming from upper and colder re- gions of the thundercloud, is cold; some of it may even be frozen. This cold deluge cools the air as it falls, partly by its own coldness, but even more by its partial evaporation. The result is that air in the heavy rainfall region is soon chilled well below the surrounding air, hence becomes denser than the surrounding air, hence begins to sink, and soon develops into a great downdraft of cold air perhaps nearly as violent as the updraft of warm air in front of it.

This cold downdraft reaches the ground with the forward speed of the whole storm at upper levels (which may be twenty or thirty miles per hour or more), and at the ground level it sweeps forward as a violent outrushing squall wind, blowing in heavy gusts directly out of the storm.

But the squall wind may far exceed the speed of the storm itself. It is not impossible for a storm moving at thirty miles an hour, say, to produce a surface squall which (however shortlived) may blow at sixty or eighty miles per hour and sweep all before it. These excessive squall winds are perhaps partly explained by the crowding of the downdraft against the ground (which also contributes to the sharp barometer rise noted with the squall).

But the chief reason for their ferocity is the action of a horizontal vortex. With a great updraft in one part of the cloud and a great downdraft behind it, all sorts of eddies and minor whirlwinds are bound to exist in thunderclouds, as airmen who have endured the blind bumpiness inside one can fervently testify.

Most of these vortices, being well up in the cloud, are invisible (except on the gyrating instrument dials of an unhappy airplane); but one of them, being formed just at the cloud base, where the relative humidity is such that the slightest drop in pressure condenses moisture, may plainly appear as the long, light-colored roll cloud or 'squall head' arching horizontally across the front of the storm.

As the roll-cloud vortex revolves, it moves across land or water like the revolving brush of a gigantic carpet-sweeper, piling its speed of rotation on top of its speed of horizontal motion — all to the consternation of squall-blown humans in one case, or insects swept out of a carpet in the other. This horizontal vortex of the roll cloud is, in fact, a sort of horizontal tornado; and under certain conditions, parts of it can turn down to the ground and become real tornadoes. In any case, it is something for an aviator to avoid at all costs.

This whole thunderstorm wind-mechanism picture, has, of course, been snapped at a single and very typical instant. Actually, thunderstorms and line squalls vary greatly in size, form, and intensity. Some thunderstorms have no squalls; some squalls have no thunder; and any one storm is constantly changing, from moment to moment, its form and intensity. Updrafts and downdrafts alike are constantly changing their speed and their position in the storm.

Thundershowers are usually of the brief, heavy type, and may reach cloudburst intensity. Raindrops in the active parts of the storm are much larger than ordinary raindrops, being mostly of near-maximum size; and they are crowded together, in certain places, by the interplay of violent and ever-varying air currents. It is not at all uncommon for these thunderstorm deluges to fall at the rate of three inches or more an hour, flooding low places with a few feet of suddenly rising water.

On occasion, the rate of water-fall for a brief interval has been known to run as high as one inch per minute! To a ground observer such a cloudburst seems to be falling in sheets of solid water, as if the heavens had indeed opened; to an aviator trying to fly through the downpour, it seems almost like boring through the depths of the ocean, and airplanes have actually been brought down by running unsuspecting into a heavy cloudburst at night.

On the ground a cloudburst is most destructive in normally dry desert and mountain regions — among steep converging slopes bare of brush or timber — where any dry ravine may within a few minutes become a raging torrent, and walls of tumultuous water ten or twenty feet high may roar along the rock-walled bottoms of steep canyons.

Nor is torrential rain the last of possible thunderstorm visitations on men caught in or below the storm. On occasion, thousands of tons of ice may fall from the skies in the destructive form of hail. As we have seen earlier in this chapter, the upper parts of a towering thundercloud harbor many slowly falling ice particles. Under favorable conditions of temperature and humidity, some of these may grow into sphero-crystals or snow balls (*graupel*) about the size of small peas.

Alternatively, or in addition, raindrops carried by the storm's great updrafts into temperatures well below freezing, and sufficiently jostled by turbulence, may be frozen into ice balls (sleet) of similar size. When the updraft becomes less violent, or when these pellets reach the edge of it, they fall groundward again, perhaps gathering a coating of snow above the freezing level, and a further coating of slush or water below it.

Soon, however, the growing ice pellet may enter new violent updrafts, which carry it upward again into the arctic zone, where its acquisitions of snow, slush, and water are frozen into a new concentric shell of solid ice, increasing its diameter. In like manner the ice pellet may fall again through several thousand feet, be carried upward again through another great rise, and repeat the whole down-up cycle ten or twenty times, growing the while into an ice ball of considerable size.

The largest hailstones ever authentically observed, which fell on some Nebraska farms in 1928, were the size of small grapefruit. One of these stones weighed one and one half pounds and was about six inches in diameter! Three-inch hailstones, the size of ordinary oranges, probably shower down somewhere in the United States a few times every summer. Stones of golf-ball size are not at all uncommon. In 1936, nineteen Bushmen in South Africa were reported to have been killed by a fall of oversized hail that covered their bodies to a depth of three feet. Hail of some sort accompanies a large proportion of temperate.

one thunderstorms, but the hail zone of any one storm is usually not more than a few miles wide. But in tropical thunderstorms hail is very rare, however heavy the lightning and rain, because rain-making portions of the thunderclouds seldom extend far into a freezing zone perhaps five miles high.

The higher a temperate-zone thunderstorm extends, the blacker are its underparts, and the more likely it is to unleash severe lightning, torrential rain, destructive winds, and hail. The convergence of two or more storms tends to intensify these factors. Wind-tunnel tests with hail-sized balls have revealed some astounding probable speeds for the updrafts of great storms.

To sustain a one-inch hailstone an updraft of about sixty miles per hour is necessary; a three-inch stone requires a hundred-mile current; and a five-inch stone must have been held aloft at some time by an upblast of two hundred miles an hour! In severe thunderstorms, therefore, 60-mile updrafts (and commensurate adjacent downdrafts) are common, while 100-mile and 200-mile currents are not impossible. Which seems good enough reason for aviators to keep out of them if they can.

An average detached or 'air-mass' thunderstorm starts as a pile of cumulus clouds perhaps five miles or so in diameter, gradually spreading out horizontally and building up vertically as it sucks in warm surface air and gathers to itself near-by clouds. The altitude of its highest cumuliform turrets may be four or five miles in temperate latitudes (or considerably higher in the tropics); and the cirriform anvil or 'cap' may reach six or seven miles aloft.

As the storm moves bodily along with the upper winds, it spreads out laterally to a width of perhaps twenty or thirty miles, so that its whole path resembles a pear in shape. After six or seven hours, with the decline and disappearance of the sun, an air-mass storm usually dies out, having raged in all through a distance of perhaps a hundred and fifty or two hundred miles.

The airplane has of course made it possible to survey the breath-taking topography of thunderstorm clouds, quickly and accurately. One of the TWA 'over-weather' research aviators made a detailed investigation of a large air-mass storm above Missouri. The general storm-top of cumuliform clouds was at twenty-two thousand feet. The highest cumulo-nimbus turrets boiled up to twenty-five thousand feet, at which altitude the plane could just clear the turbulent wind-shift line.

The cirriform anvil, ten miles wide, extended on up to about thirty-five thousand feet. Flying through the anvil at twenty-six thousand feet, the plane encountered fine snow and some subcooled water droplets which froze as light icing; the temperature was 5° F., compared with 100° F. at the ground.

Even an ordinary thunderstorm has within it a cold front in miniature — a rain-cooled wedge of squall-wind air surging outward and displacing the warm air in front of the storm. But the cold air behind a large-scale front usually comes from arctic or polar regions, whereas the cold air in a thunderstorm is, one might say, manufactured locally by the refrigerating process of the storm. Yet if a very large thunderstorm spreads out laterally beyond fifty miles or so, it begins to take on the characteristics of a line squall.

THE WINGS OF THE STORM

Reduced to typical form and simplest terms, a line squall is nothing more than a violent cold front often spotted with many thunderstorms, continuous or intermittent, along the hundreds of miles of its length. The front itself is necessarily marked by more than ordinarily large temperature contrast between the unlike air masses; by high moisture content and great depth of the warm air mass; and perhaps also by some moisture and instability in the cold air mass.

Also in evidence is the rapid advance of the cold air mass, the enforced retreat (mostly sidewise) of the warm air mass, and conflicting impact between the two. This air-mass conflict shows up as high winds along and across the front, sharply contrasted in direction, and appears on the weather map as isobars packed close together and bending in a sharp V where they cross the front.

From this typical picture as a norm, individual line squalls of course vary widely. Few cold fronts are violent enough to be spectacular line squalls, and few of these, even, conform strictly to type. The one common characteristic of all surface cold fronts, at least over flat country or the ocean, is a marked shift in wind direction, usually from some southerly to some northerly quarter.

The squall line itself usually advances from the northwest or west (less frequently from the southwest) at something like twenty to thirty miles per hour in summer and thirty to forty miles per hour in winter. (Occasionally, in some parts of the country, the squall may come out of the north or even the northeast.)

But rather than further generalities, often violated, here are accounts of two actual line squalls that struck the city of Washington. One came early in June: At 2 P.M. the sky was still clear, but hazy; and a few clouds were rising in the distant west. The wind was moderate out of the southwest, and some twenty or thirty small sailboats were racing on the wide Potomac River, which was also dotted with various other craft. Upper clouds gradually grew out of the west, gradually spread and darkened, obscuring the sun in a quarter of an hour.

Then a storm collar or squall cloud appeared at lower levels out of the southwest. It consisted of dark gray stratocumulus clouds, whose violent agitation showed the ferocity of their inner winds.

The squall cloud came rapidly on until it passed overhead, but there was no lightning and no rain. The wind slackened somewhat directly under the squall cloud, becoming rather puffy, and the sailboats continued their gay progress all unheeding. But perhaps five minutes after the squall cloud had passed overhead, the surface squall struck very suddenly and fiercely out of the southwest.

As the wind rose shrieking to fifty or sixty miles per hour the sailboats, one after another, were pressed down to the water and capsized, some with sails still set, some under bare poles. The extreme wind lasted only three or four minutes, then moderated to twenty miles an hour for another half-hour, with light rain; after which it fell nearly calm while the rain continued.

Another squall came in May, when the late spring contrasts between tropical and polar air were most intense:

Up to mid-afternoon the sky was perfectly clear but very hazy, the wind moderate southwest. About 3 P.M. the forerunner clouds appeared out of the northwest — perhaps a sort of upper-level squall head — arching across the sky from northeast to southwest. They appeared as a solid wedge of high strato-cumulus or low altocumulus, advancing steadily at an altitude between one and two miles.

There were moderate puffs of wind out of the northwest, followed by a gusty northwest wind of twenty-five or thirty miles per hour — but for the moment, no rain. The high clouds swept by overhead and out to eastward, shutting out the sky, and for twenty minutes nothing more happened, except that the pressure jumped up by five hundredths as the cloud edge passed, and the gloom steadily deepened in the northwest. Then the gray rain curtain appeared, and with it came the real squall at 3.50 P.M. For a few minutes the northwest wind whistled around sixty-five miles an hour, driving torrential rain and rattling hail horizontally against windows and doors. Lightning flickered incessantly above and banged heavily to ground, trees and poles were borne to earth in exposed places, and the horizontal rain flooded over everything, cutting visibility to zero.

The base of the black thunderclouds, when they could be seen at all, looked about one thousand or two thousand feet high. The temperature fell twenty-four degrees in an hour. The worst of the squall was over in a few minutes, but scattered showers and mild thunderstorms continued, interspersed with broken clouds, throughout most of the evening in the unstable invading cold air mass.

These are the picturesque and transient line squalls of spring and summer. In winter, marked cold fronts are less spectacular, but if anything more unpleasant. Their approach is more rapid, and the cold air mass is much colder. Intense winter fronts are likely to be marked by a deep zone of blizzard-like snow.

The worst weight and hazard of summer line squalls comes on the water, as the sailboat episode above indicates.

FLYING THROUGH A LINE SQUALL

Full-bodied line squalls are colorful enough, viewed from the ground or the water. But the way to see and feel one at its best — or its worst — is to get up into the air where we are almost a part of the gyrating squall winds.

This is something that wise aviators seldom do, intentionally (except in big, scheduled transports, or in urgent military flying); but they sometimes get caught in squalls or plow through them deliberately in the course of an urgent mission. On the same afternoon that the airship *Hindenburg* crashed in flames at Lakehurst, an army pilot and I were flying from Boston to Washington in *BT*-8 — a medium-fast, stubby, powerful pursuit-type training monoplane:

Towards Trenton the clouds were more scattered, and we dropped down through them to 2000, heading out westward so as to pass well to windward of Philadelphia's smoke pall. The scattered cumulus clouds, now above us, were becoming larger, with some scattered showers among them. Some of these shining white cumulus heads were beginning to mushroom quite rapidly up into the blue — a sign of thundery weather.

Soon there appeared far ahead, somewhere over northeastern Maryland or southeastern Pennsylvania, a dark line or roll of cloud arching across the entire sky from southwest to northeast. This was the real front, the line squall itself. Great white thunderheads towered along it, in some places much higher than others. Beneath these higher cloud-peaks were gray curtains of rain, and around them flickered the seeming-miniature sparks of lightning. For the time being, the controls were in my not very experienced hands.

Our course was about 240° magnetic (southwest-by-west true), and as we approached to within ten miles or so of the squall line, made out two large thunderheads towering far above the long, continuous roll cloud — turreted white mountains three or four miles high. Under each thunderhead a dark deluge of rain was falling from the squall cloud to the ground.

It seemed wise to keep out of these active storms if possible, and we changed course 15° to the right so as to fly midway between them, just under the long, black roll cloud (which appeared fortunately to be well up off the ground, at a height of half a mile or more).

Beneath this rapidly approaching black roof of turbulent cloud the visibility was good — we could see far out ahead into patches of sunlight and shadow beneath other clouds behind the front. We were still cruising in level flight at 2000, and the black ceiling at about 3000, now rushing towards us overhead, seemed to offer plenty of clearance.

As we passed beneath the dark cloud-roof the *BT*-8 began to bounce around crazily. I had expected bumps, but these were so sharp and ferocious that it was all do to maintain oncourse control of a plane that was, at best, rather tricky to fly. Snatching a glance at the instruments, was alarmed to see the rate-of-climb meter swing to plus 1000 ft/min (12. m.p.h. straight upward), in spite of the fact that was already nosing downward in an effort to get lower and escape from the wildly gyrating black festoons of cloud that seemed now just over our heads.

This black cloud-roof, marked by turbulent downward bulges of mamato-cumulus and momentary 'cloud spouts' or incipient whirlwinds, was getting nearer by the second. It seemed as though we were about to get sucked bodily up into its black chaos. I had the nose shoved down as steeply as seemed prudent, for the air speed was now over 200 m.p.h. and the terrific bumps might strain the wings at this speed. Yet I hesitated to change the throttle setting without instructions from the pilot.

Finally I jammed the nose down in a steeper dive, taking a terrific beating from the bumps and still not losing altitude. At this point the great down-bulging cloud-pocks seemed to be almost couching our heads. Just then the pilot took over, and throttled back to case the bump-strains. And within another minute, we had passed out from under the roll cloud into a region of scattered sunlight and shadow, marked by detached storms in the cold air mass. These detached storms, of course, could be easily flown around.

Passing through a line squall in a controllable airplane may be exciting enough, but it is relatively easy and pleasant compared to navigating through one in an uncontrollable free balloon. Lieutenant Settle of the United States Navy had this unenviable experience over western Pennsylvania on the twentyeighth of May, 1938. Wrote he, in part:

Dropped ballast and at 18.15 (6.15 P.M.) balloon was at 4500 feet altitude; the line cloud overtook and passed under us and we started up rapidly, the variometer going 'hard over' on the ascent side. At 8300 feet I valved two seconds, and balloon leveled off at 8400 feet and then started down at a violent rate, the variometer ink going out of sight on the descent side. While falling the air was extremely turbulent; hail (size of peas), heavy rain, and snow flurries hit us; the basket swung violently from side to side like a pendulum and spun in azimuth; heavy gusts hit the envelope and basket; the motion of the balloon was like when attempting to hold one on the ground in very gusty surface wind. Lightning and thunder were heavy and apparently right on top of us. The rain and air were very cold and our hands numbed.

During the descent we dropped approximately fifteen bags of

ballast; we came through the clouds at 3000 feet and leveled off at about 2000 feet. Found myself under a dense black cloud, the line boundary of which was several miles south of me; the line extended as far as see in either direction. There were areas of heavy rain along the line on both sides of me and heavy rain to the northward of me, with lightning and thunder on all sides. The rain in my immediate vicinity had temporarily ceased; my air was NNW, about thirty miles per hour, and for a few minutes seemed relatively smooth.

It was soon apparent that we were fast overhauling the line cloud. When nearly up to the line, we again started violently up through the cloud on a second 'vertical circle,' repeating the history of the first one, and attaining a maximum altitude of 8200 feet. Some of the elements were more violent in the second than in the first excursion. Settle finally managed, by adroit maneuvering with gas valve and ballast release, to case his balloon out of the squall line. He landed at dusk on a farm near Parryopolis. He lived to write of his experience, but others who dared line squalls in free balloons, such as the brilliant American aerologist Meisinger, were less fortunate and plunged to lightning-lit deaths.

Chapter 10

Movement of Earth Crust

The miner deep in an Illinois coal mine who unearths a tree stump with the roots spreading out in the position in which they grew concludes without question that this tree, now 1,000 feet underground and 600 feet below sea level, once grew on the surface of the earth. Similarly, the manager of a copper mine in northern Michigan whose ore comes from typical stream-deposited conglomerate or from the scoriaceous upper part of a basalt flow a mile underground reaches the same conclusion.

The thought that land and sea are not always stationary but must locally have changed places has been entertained by nearly all peoples who have attempted to explain the occurrence of well-preserved marine fossils in the rock of plains, deserts, and mountains. How else can we explain the marine fossils in the rocks of the Paris Basin, and the deviation of the stratification in these rocks from the horizontal?

DISPLACEMENTS OF THE EARTH'S CRUST DURING EARTHQUAKES

Spectacular but relatively small earth movements have accompanied many earthquakes Therein are described several destructive earthquakes that devastated densely populated areas. Most of these destructive earthquakes caused fissures to open in the ground and visible displacement of the earth's crust along the fissures. Such displacements are seen in the severed ends of roads, fences, strata, or other features that had extended uninterruptedly across the site of the break prior to the earthquake.

Breaks in the earth's crust along which slipping has occurred are called *faults*. Fault movement does not always produce an earthquake; slow, gradual movements along some faults have been observed without accompanying earthquake shocks. Sudden displacement along faults, however, appears to be the chief cause of earthquakes.

The fault displacements observed during severe historic earthquakes are small—seldom more than a few feet. In some, the movement was vertical, producing a small cliff along the fault; in others (for example, the destructive San Francisco earthquake of 1906), the fault walls slipped laterally, offsetting roads or fences that formerly crossed the break; and in still others (for example, the Mino-Owari earthquake in Japan), the slipping was oblique, with both vertical and horizontal components.

Perhaps some historic earthquakes have produced even greater fault displacements on the sea floor. After an earthquake near Disenchantment Bay, Alaska, in 1899, the beaches stood 47 feet above the sea, and a wide expanse of sea floor was made into dry land. This is the greatest well-authenticated single displacement known to have accompanied an earthquake.

The great Japanese earthquake of September 1, 1923, which largely destroyed Yokohama and Tokio with a loss of over 140,000 lives, was at first believed to have been accompanied by a much greater displacement, but the evidence is inconclusive. Comparison of soundings in Sagami Bay made before and after the earthquake showed local changes of more than 1,000 feet in depth of water. Detailed study suggests, however, that most and perhaps all of these great changes in the sea floor were not from fault displacement but more likely from large slides set up in the unconsolidated muds and silts on the floor of the Bay.

Thus, the historic record indicates that no single observed fault displacement accompanying an earthquake has been drastic enough to account for marine strata on mountain tops or rooted stumps in mines 1,000 feet or more below sea level. Is it possible, however, that these small displacements are but minor episodes in the growth of the fault on which they occur? Examination of the structure of the rocks on either side of many earthquake fissures affirms this question.

Although new faults with a few feet of displacement have formed during a few recorded earthquakes, most displacements have taken place by renewed movement along an old fault. The presence of such a fault is revealed by the impossibility of matching strata or other structures in the rocks on the two sides of the fault by merely restoring them to the positions they occupied prior to the earthquake.

The Pleasant Valley earthquake, which rocked the almost uninhabited desert of central Nevada in 1915, offers a good example. After the earthquake, the western base of the Sonoma Range was marked for 17 miles by a low cliff 1 to 16 feet high where none had existed before. The cliff was formed by renewed movement on an old fault which bounds the low-lying, partially alluvium-covered Pleasant Valley on the west from the rugged Sonoma Range to the east. During the earthquake, the Sonoma Range, relative to Pleasant Valley, moved directly up the steep fault at its base, forming the new cliff and increasing the height of the range up to 16 feet.

Displacement is clearly seen in the vertical cliff offsetting the surface of the alluvium. The small patch of alluvium clinging to the surface of the upthrown Sonoma Range block is 12 feet higher than the alluvial surface on the Pleasant Valley block. But the total displacement on the fault must be much greater; the dolomite of the upthrown block is against alluvium on the downthrown block for a vertical distance far greater than 12 feet.

Furthermore, a short distance away, gullies cut in the Pleasant Valley alluvium show that it rests not on dolomite like that across the fault, but upon lava flows. Immediately across the fault, however, dolomite and other sedimentary rocks extend to the crest of the Sonoma Range over 2,000 feet higher. Still farther north, the higher peaks of the Range are capped by lava flows like those on the Valley floor.

Here, however, the lavas rest on the dolomite and associated sedimentary rocks. Obviously, therefore, the total vertical displacement on the fault has been at least 2,000 feet, and perhaps it has been enough to account for the entire difference in level between valley and range, which is locally more than 3,500 feet. Apparently, the 1915 displacement was only the latest of many similar movements.

Other historic ground displacements that have accompanied earthquakes show similar relations and warrant the conclusion that repeated small movements along a fault can ultimately displace the crust for thousands of feet.

Measurable Slow Movements Along Faults

The Buena Vista oil field is on a rounded hill rising above the flat San Joaquin Valley in California. Soon after the first oil wells were drilled, a road was built across the hill, and several pipelines were buried in shallow ditches along it. Within a few months, the roadbed cracked, and a vertical displacement of less than an inch appeared on its surface.

It was patched, but repeatedly broke at the same place. Within a few years, eight of the buried pipelines had buckled up out of the ground, and they continued to arch upward slowly, a few inches each year. The pipelines were thus shortened 9 to 19 inches in a period of 9 to 15 years.

Cleaning tools lowered into the wells also revealed that 23 of the well casings were slowly being bent out of line. In some wells, bending continued until the casing col lapsed; in others, the casing, when pulled from the ground, was found to have been displaced about 15 inches horizontally within the 10 to 15 years after the well had been drilled. The casing failures in different wells occurred at depths ranging from 76 feet to 794 feet. What caused these curious phenomena?

By plotting on maps and sections all of the points where such disruptions occurred, it was found that they lie along a smoothly curving surface. This surface slopes downward into the ground at a low angle, and the rocks it cuts are offset against it. Both the disturbances of man-made structures and the offsets in the rocks prove that the surface is a fault.

Apparently the rocks above the fault surface are moving slowly upward and southward with respect to those below, breaking the roadbed, bending the well casings, and buckling the pipelines out of the ground. Surveys show that the average rate of movement on the fault is 1 1/2 inches per year: No earthquake shocks arising from this fault have been detected, even by seismographs.

Other examples of slow continuous movement along faults are known. In some parts of Japan, differential movements between different blocks of the earth's crust separated by faults have been detected by resurveys of the position and elevation of a series of points established by precise triangulation and leveling. For example, there were changes in level of a series of points along a 150-kilometre stretch of the northeast coast of the island of Honshu that occurred between leveling surveys made in 1900 and 1933.

The changes indicate that throughout this area the earth's crust sank relative to sea level. The amount of displacement varies at different points, supposedly due to slow tilting and warping of fault-bounded blocks in the earth's crust.

Measurable Slow Movements Not Connected with Faults

The movements have taken place along faults, but there are also widespread and important movements in no way related to faults. Although the rate of movement is slow, it has been measured at several localities. In a few places such movements have changed the landscape appreciably within historic time.

The classical and perhaps best known example of such movements is that of an ancient Roman ruin, the so-called "Temple of Jupiter Serapis" near Naples, Italy. Only three upright columns and a part of the floor are still in place. On the columns about 18 feet above the floor is a line. Above this the columns are smooth; just below it they have been bored full of holes by a marine rock-boring clam.

Shells of this clam can still be seen in some of the holes. It is reasoned that the temple was built on dry land, that slow downward movements then brought sea level to a point 18 feet above the floor, allowing the marine clams to bore the columns, and that, later, the land rose again.

Along the shores of the Baltic Sea in Sweden and Finland a still more striking movement has been observed for many years. The farmland and fresh-water marshes fringing the Baltic are littered with marine shells identical with those in the Baltic today. Over 150 years ago these conditions were correctly interpreted to mean that the land had risen out of the sea. In order to find out whether the movements are still going on, monuments were set up at high tide line along the shore.

Today many of these monuments are a few feet above sea level, and some as much as a mile inland. The maximum rate of uplift—3 to 4 feet per century—is in the northern Baltic. In some parts of Scandinavia there is little or no movement, and southern Denmark appears to be sinking about 2 feet per century. The rate of movement also appears to vary somewhat with time, as considerable differences from one decade to the next have been revealed by accurate tide gage records at many Baltic ports.

What causes this uplift? When the measured changes along the Scandinavian coast are compared with the obvious prehistoric evidence of uplift from raised marine beaches and terraces, an interesting relation appears. During the Pleistocene the Scandinavian area was covered by a huge sheet of ice such as mantles Greenland today. The Scandinavian uplift is greatest in the area where the Pleistocene glacier was thickest. The ice sheet was thickest in the northern Baltic, and in the approximately 12,000 years since the ice melted (dated by varved clays in glacial lakes.

Because of the tendency of the earth's crust to reach isostatic equilibrium, uplift is what we would expect from unloading of a segment of the earth's crust by the melting of a large ice sheet, and the greatest uplift should take place where the greatest load was removed. Thus, this close coincidence between ice thickness and uplift is strong geologic evidence of the isostatic tendency. Since the uplift is still going on, we infer that Scandinavia has not yet risen to the elevation it must have had before the load of glacial ice began to accumulate and to depress the earth's crust under its weight.

Yet, there was no glacier or other large load upon the land at the Temple of Jupiter Serapis, which also sank and later rose. There must, then, be forces other than those tending to restore isostatic balance which are involved in crustal movements.

GEOLOGIC EVIDENCE OF DISPLACEMENTS OF THE EARTH'S CRUST

Emergence and Submergence Along Coasts

Along most of the world's coastlines there is evidence of geologically recent (though prehistoric) crustal movement. The lobster fisherman who sets his traps in the water off the Maine coast now and then brings up fragments of the peat that floors the bottoms of some of the inlets and bays. In this part of the United States, almost every stream ends in a tidal estuary. Dredgings from the bottom of these estuaries usually reveal river-deposited silts filled with decomposed grass roots, peat, or thinly stratified clay formed in freshwater lakes and marshes. All of these must have been deposited above sea level.

On a clear day, a person flying over southern San Francisco Bay can see the drainage channels of a former land on the shallow Bay floor. These channels, though interrupted by deltas or small wave-cut features at the shore, are clearly underwater continuations of the streams now entering the bay.

They have apparently been drowned either by subsidence of the land on which they were flowing, or by a rise of the sea. Not far away, near Stockton, California, water wells drilled to as much as 1,000 feet below sea level penetrate buried soils, river silts with grass roots, peat accumulated in fresh-water marshes, and other materials that must have been deposited above sea level. Off Denmark and Japan, dredging in shallow marine water has revealed kitchen middens (refuse heaps left by primitive man) and accumulations of charcoal and ashes marking the position of ancient campfires.

High above the reach of present-day waves, along the shores of Alaska, Labrador, Newfoundland, Chile, the East Indies, Scandinavia, Italy, and many other countries, are abundant relics of former shorelines. Such features include barnacle shells and coral still attached to the rocks in the position in which they grew; rocks bored by marine clams (many with the shells still in the holes); sea cliffs, sea caves, stacks and other erosional features carved from solid rocks by former waves; and deposits of shell-strewn sand drifted along the ancient beaches by longshore currents.

Conspicuous marine terraces border many sea coasts. Although locally masked by detritus washed down from the hills, or partly cut away by streams since their uplift, most of them preserve shellfilled sands or other clear proof that they are the uplifted floors of ancient seas.

Some of the most diverse examples of geologically recent changes of level are found in the East Indies, particularly in the region north of Sumatra, Java, and the southern Moluccas. Fringing coral reefs abound in the warm clear seas near these tropical lands. Growing reefs generally have relatively flat tops with surfaces less than 150 feet below sea level.

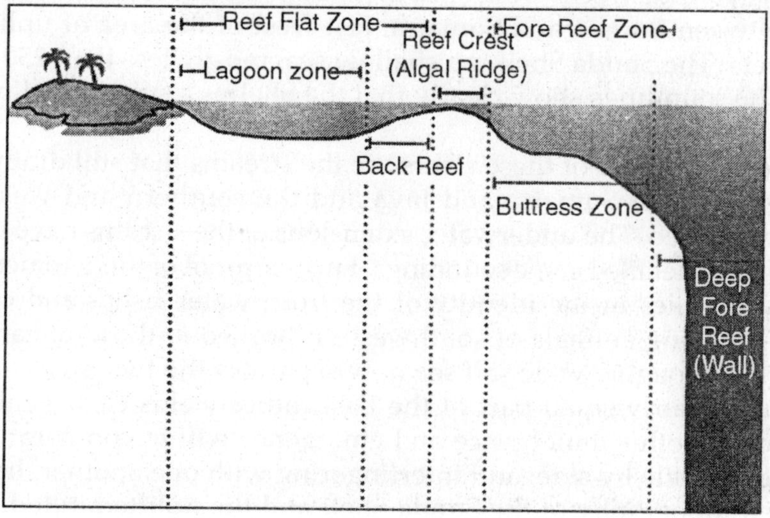

Fig. Diagrammatic Sketch of a Typical Coral Reef.

The reef-building coral animals cannot stand depths greater than 200 feet and are also killed by a few hours' emergence. Coral limestone, largely built of reef corals and associated organisms, is, however, by no means confined to the present zone of coral growth. Far above the reach of the highest storm waves are great terraces of white coralline rock exactly like that accumulating off the present shore. Some of these uplifted reefs are only a few feet above the sea, others fringe mountain tops 3,000 feet high.

Some uplifted reefs form continuous level collars around the smaller islands. Others have been tilted, for they may be 100 feet or more above sea level on one side of an island, but when traced around to the other side they gradually decline to sea level, or even pass below it. Still other uplifted reefs have obviously been faulted and warped.

Each of these raised reefs, of course, testifies to uplift of the land with respect to the sea. But, as if to contradict this, we find equally clear evidence that some former land areas now lie depressed beneath the ocean. The chief evidence is from two sources: examination of samples dredged from the sea bottom, and fairly detailed contour maps of the ocean floor prepared from sonic soundings. Though coral reefs grow only in comparatively shallow water, at various points in the

East Indies the dredge brings dead reef corals from depths of more than 1,000 feet. It also brings up land deposits such as muds that accumulated in fresh-water marshes and mangrove swamps or the poorly sorted silts and sands typical of river floodplains.

Sonic soundings also show indubitable evidence that some of the shallow East Indian seas were once dry land. One example is the Sunda Shelf between Borneo and Sumatra, just west of the area of uplifted coral reefs. The Sunda Shelf is a shallow sea, mostly less than 250 feet deep. The soundings show clearly that the Shelf is a submerged plain with three trunk rivers.

The headwaters of these rivers are the streams that still drain the northern parts of Sumatra and Java and the southern and western slopes of Borneo. The underwater extensions of the streams have been located from detailed sonic soundings. Further proof of this submerged river system lies in the identity of the fresh-water fishes and other stream-dwelling animals of southwestern Borneo to those of eastern Sumatra, although a wide salt sea now separates the two areas.

Thus, a fairly small part of the East Indies yields clear geologic evidence of both submergence and emergence within comparatively recent time. Side by side, and interfingering with one another, lie the drowned topography of the Sunda Shelf and the uplifted, tilted, and faulted coral reefs of the southern Moluccas. Elevated reefs are also found in Java and Sumatra, pointing to early submergence and then uplift prior to the drowning of the river system on the Sunda Shelf.

Obviously, the movements that have changed the levels here have not been simple. They involve not merely vertical uplift and subsidence, but concurrent bending and folding of the earth's crust. Indeed, many of the tilted reefs fit into a regional pattern of folds in the basement rocks beneath the surface. Study of the rock structure of the islands shows that most of the long curving island arcs are complex upfolds or arches in the rock, separated by basins and downfolds that still lie, in general, below the sea.

The movements that produced these upfolds and basins are still going on, warping the fringing coral reefs above the sea at one point, drowning the reefs and the land-laid deposits behind them deeply at another, and forming a complex but slowly growing system of folds in the crust of the earth.

Contemporaneous Folding, Erosion, and Deposition

The folding of the earth's crust disclosed by the uplifts and submergences in the East Indies suggests that possibly such movements are related to the formation of mountains. Many mountain ranges show warped and bent strata that must have been deposited as nearly horizontal sheets.

They are now distorted into patterns more complex than the East Indian warps, but suggestive of them. Can we find any geological links between the broad warping of the East Indies and the more intense localized folding of mountain ranges such as the Alps or the Appalachians? Let us consider first the relatively simple warps and folds in the lava flows that make up the eastern foothills of the Cascade Mountains in northwestern United States.

In central Washington and northern Oregon the walls of the canyons of Columbia River and its tributaries reveal flow upon flow of black basalt. The flows are 50 to 400 feet thick, and, unlike those at Stolpen near Freiberg, have relatively little interbedded sedimentary material. Here and there, however, thin sheets of river gravel separate the flows, showing that streams advanced onto the barren volcanic plain after some eruptions only to be overwhelmed by new lava flows.

Elsewhere, narrow bands of red or black soil containing petrified (silicified) logs, or tree stumps with their roots still spreading out in the soil, appear between the flows. Obviously, time enough had elapsed for the scoriaceous flow tops to weather into soil and for a forest to grow on it before a new sheet of lava devastated the area. Elsewhere, layers of a brilliant white rock with paper-thin stratification contrast strikingly with the basalt.

Petrographic examination shows that these white rocks are made up almost entirely of the siliceous skeletons of minute onecelled plants called *diatoms*. Similar white diatomaceous deposits are being laid down today on the floors of shallow ponds and lakes nearby. In one part of central Washington, river gravels, floodplain silts, and diatomaceous deposits continued to accumulate on the surface of the lava plain long after volcanism had ceased, until locally the black basalt flows were buried under several hundred feet of these unconsolidated river and lake deposits.

The point to be emphasized is that at many places these basalt flows and associated sedimentary rocks are no longer horizontal. They rise and fall in great arches and troughs. Here, on the flank of an arch, the scoriaceous top of a basalt flow and the diatomite strata above it dip at an angle of 20 degrees. There, beds of gravel and sand between basalt flows dip at 70 to 85 degrees. Obviously, no stream tumbling down a 70-degree slope could deposit gravel and sand upon it, nor could a molten basalt flow drape itself over a steep-sided arch and maintain a uniform thickness.

Nor could the paper-thin sheets of diatomite remain on a 20degree slope on a lake bottom. They must originally have been nearly horizontal. Clearly, the flows of basalt and sheets of gravel and diatomite were folded into the present arches and troughs after they had been deposited.

What is the relation of these arches and troughs to the present topography? Central Washington is rather exceptional among mountainous areas in that the major topographic ridges coincide with the arches in the lava. The flanks of the arches are scored by many ravines, nearly parallel, extending straight down the slopes. Turbulent floods course down these ravines in the wet season, cutting deep canyons into the sides of the arches.

Generally, these streams, aided by rillwash and downslope movements, have removed most of the unconsolidated river and lake deposits from the tops and flanks of the arches, but the eroded edges of these unconsolidated beds can still be seen farther down the flanks of the folds. In some places, a few of the much more resistant basalt flows have also been eroded, and locally so many of them as to erase almost completely the fold as a topographic feature.

In general, however, the smooth curving surfaces of the arch, intact except for small ravines, indicate that erosion is only getting started in demolishing the uplift. Many of these ravines do not have the typical concave-upward profiles of normal streams. The stream canyons and the folds have grown synchronously, and slow continuous arching of the lava sheets has modified the development of a normal stream regimen.

What is the relation of these topographic ridges to the troughs alongside? Conditions vary locally, but some of the partly dissected arches lie half buried in unconsolidated sediments deposited by streams and in lakes on the floors of the troughs. The uppermost beds of these trough deposits are nearly horizontal, abutting against the tilted rocks of the arches at high angles. In a few places, younger lava flows are interbedded with these young sediments, but they do not extend onto the arches. They were erupted after the folding had started, and poured down the troughs.

In a few of the larger river canyons we can see, and elsewhere from records of wells drilled in the troughs we can infer, that the older beds of even the younger strata have also been somewhat folded. Where they abut against the neighboring basalt arch, they dip with the basalt, though at a lower angle. Beds higher in the unconsolidated series dip similarly, but at still lower angles. In some troughs only the uppermost beds are nearly horizontal.

These relations confirm our conclusion from studying the erosion of the arches: While the troughs were being downfolded, sedimentation was also going on. In a few places, sediments were deposited on the trough floors almost as fast as they grew, and have been progressively warped as the fold developed.

Some of the sediments filling these troughs were eroded from the adjoining growing arches, but much was brought in by rivers from the rugged Cascade Mountains to the west, by the abundant quartz, pumice, and andesite fragments—constituents foreign to the adjacent arches.

Such evidence of contemporaneous erosion and sedimentation must mean that the folds grew very slowly. Individual flows can be traced from the crest of an arch down its flank until they disappear beneath the unconsolidated sediments of the adjacent trough; but across the basin they reappear and can be seen to rise in the next arch. Thus, originally horizontal flows have been folded until the surface of a particular flow on the crest of an arch is thousands of feet higher than it is at the bottom of an adjoining trough. But the folds did not grow overnight, for it is clear that folding, erosion, and sedimentation were all occurring simultaneously.

The first stage of the folding blocked out the respective areas of erosion (arches) and sedimentation (troughs). The rising folds dammed streams and impounded lakes in the troughs, interfering with normal stream development. Sedimentation also was locally interrupted by tilting of already deposited beds. The concurrent erosion of the arches and deposition in the troughs as the folds grew shows that the movements were slow and gradual. There is meager but not wholly conclusive evidence from changes in the speed with which water is reported to flow in some of the irrigation canals built 10 to 50 years ago that the folding movements are still continuing, and thus changing the gradient of the canals.

We could cite many other examples to show that warping and folding go on slowly along with erosion and deposition. This has indeed been the rule in the young and growing mountain ranges fringing the Pacific Ocean, and in the Alpine-Himalayan belt across southern Eurasia.

FOLDS THAT HAVE CEASED TO GROW

An older mountain range such as the Appalachian offers both contrasts and similarities to the structures of the East Indies and of the Cascade foothills cross section of the folded rocks of a part of the Appalachians of central Pennsylvania. This section depicts in detail the position and attitude of the rocks deep underground as compiled from surveys of the extensive coal mines and many exploratory drill holes. The independence between the folds of this section and the topography suggests immediately that the Pennsylvania folds are not growing today, nor have they in the recent past, as in the East Indies and the Cascade foothills. The folds have obviously long been dead, for the arches are no longer ridges nor are the troughs receiving sediments. In fact, throughout the folded Appalachians, it is common to find erosion-resistant rocks along the axis of a downfold forming a mountain summit, whereas adjacent upfolds with more easily erodible rock cores have been eroded to lowlands.

In the Appalachians, the topography is indeed closely adjusted to the rock structure, but the adjustment is of a kind quite different from that of the East Indies and the Cascade foothills. In the Appalachians, all of the ridges are underlain by erosion-resistant rocks and the lowlands by easily eroded rocks. Stream erosion, uninterrupted by folding and warping, has had time to etch out the easily eroded rocks and leave the resistant ones standing in relief.

This contrasts with the East Indies and central Washington, where soft sedimentary rocks often cap the very summit of a mountain. Clearly, the arches and troughs in Pennsylvania are not growing today, nor have they grown in recent geologic time.

Can slow crustal movements that took place in the far-distant geologic past, long before the present streams developed their courses, explain the folded beds of the Pennsylvania coal mines? The evidence is certainly less clear. Obviously, erosion has been much deeper in Pennsylvania than in central Washington or the East Indies. Gone are the sediments, if any, that were deposited in the troughs while the folds were growing.

Any emerged shoreline features that may once have clung to the initially rising arches nave long since been eroded away. We can easily measure the fault displacements shown by the offset of the individual coal beds in the section, but we cannot tell whether they required a few seconds or millions of years to form.

However, nothing in the Appalachian structure differs fundamentally in form from that of the East Indian arches and troughs. Although the underground form of the East Indian folds is concealed, their surface form and the canyon exposures suggest that the folds extend to great depths.

In the higher and more rugged parts of the Cascades to the northwest of the foothills area, erosion has already torn down some of the basalt arches and stripped so much sediment from the intervening troughs that the best evidence for their slow growth has here disappeared. Thus, there is nothing about the Pennsylvania folds requiring different conditions of growth. Slow movements of the crust, like those now proceeding in the Dutch East Indies and Japan, are entirely adequate to explain the folds of Pennsylvania as well as those of the Cascade foothills.

"The present is the key to the past." Fossil marine shells at the tops of mountains and soils and fossil plants in deep mines are but what we should expect and predict from the evidence of slow but measurable changes of level that we can see taking place in different parts of the world today.Furthermore, we can observe that, in some places and at certain times, the earth's crust has broken much like a brittle rock might be expected to break—along sharply defined faults.

Elsewhere, and at different times, in the very areas of faulting the rocks have folded like plastic dough. The reasons for this differing behaviour are obscure, for both kinds of deformation have taken place at the very surface of the ground. Nor is the ultimate cause of the deformation yet understood.

RECORDS OF EARTH MOVEMENTS

There are constructional processes to counteract the effects of erosion and so to prevent reduction of the lands to the level of the sea. One such constructional process is *volcanism*—the effusion of molten rock from the earth and its piling up on the surface.

We have also seen that in places the surface of the earth is moving up or down with respect to sea level. A second major constructional process, then, is the slow gradual warping and uplift of segments of the earth's crust.

"ORIGINAL HORIZONTALITY"—THE KEY TO STRUCTURE

How can we analyse this interplay between destructional (erosional) processes and constructional processes such as volcanism and uplift? We have seen that by the middle of the Seventeenth Century Nicolas Steno had worked out the "Law of Original Horizontality"—the proof that strata are deposited in nearly horizontal layers, parallel or nearly parallel to the surface on which they accumulated. But another century passed before geologists began using Steno's law to work out systematically the significance of the structural features found in the stratified rocks.

In lowland areas large-scale structural features such as folds and faults are generally masked by soil. It is to the mountains and deserts that the geologist must go to find large rock structures clearly displayed. Until late in the Eighteenth Century, however, the mountain areas of Europe were little studied because of the difficulties of transport and the fear of bandits.

Horace Benedict de Saussure, a Swiss geologist who was also an ardent mountain climber, did much to arouse the modern spirit of mountaineering and to develop interest in structural geology.

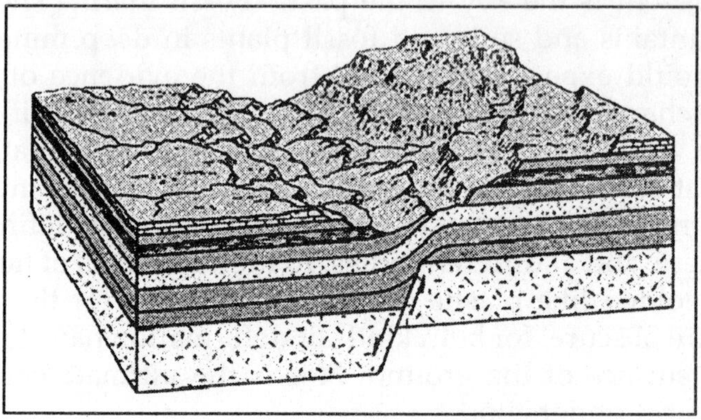

Fig. Folded Sedimentary Beds.

While climbing in the Alps, de Saussure noted that the strata appeared to be crumpled into folds like a rug pushed together in a heap on the floor, but he did not immediately grasp the significance of the law advanced by Steno more than 100 years before. Instead, he at first assumed that the rocks had crystallized in the forms we now find them. It was not until he had carefully studied beds of conglomerate projecting vertically from the ground that he recognized it was absurd to believe that sheets of pebbles, interlayered with sand and all standing on end, could ever have been deposited in that position.

Since de Saussure's time generations of geologists have scaled the peaks of the Alps and those of most other mountain ranges and have plotted on maps and sections the geometry of the deformed strata they found. Aided by what has been learned from the structural features so well displayed in mountain cliffs and canyon walls, they have also been able to work out the underground structure of plains and lowlands, despite the extensive soil cover.

The key that unlocks the geometry of these structures, and through this the sequence in which they were formed, is the principle, proved by Steno more than 200 years ago, that stratified rocks were originally nearly horizontal.

WARPS AND GENTLY TILTED STRATA

Over much of the earth the sedimentary strata appear to be still horizontal. If, however, we determine the elevations of several widely separated points on the upper surface of a particular bed, we generally find that the horizontality is not perfect. The bed either had an originally sloping surface or it has been slightly warped. Commonly, the gently inclined surface of the bed rises and falls so irregularly that its departures from horizontality cannot be explained entirely in terms of *original dip*, and so the surface must have been warped.

The rocks of the Mississippi Valley afford good examples of such broad warps. In small exposures the beds appear flat, but if we determine elevations on the upper surface of some widespread formation we find notable differences. For example, in wells drilled for petroleum in southern Illinois the Chattanooga black shale of Devonian age lies about 4,500 feet below sea level.

When followed eastward it rises steadily until, finally, it appears at the surface about 500 feet above sea level near Louisville, Kentucky. It rises similarly, though less steeply, to the north, south, and west. Well records and mapping reveal that the top of the Dakota sandstone, a widespread formation of Cretaceous age, is about 3,000 feet below sea level near Williston, North Dakota, but is 2,000 feet or more above sea level near Shelby, Montana, and 1,000 feet above sea level near Sioux City, Iowa.

Such broad warps may pass gradually into folds—the Chattanooga shale is highly folded in the Appalachian Mountains, and the Dakota sandstone has been wrinkled into folds and broken by faults in the Rockies—but the characteristic feature of a broad warp is that the beds are still nearly horizontal, even though we find appreciable differences in elevation when large areas are considered. Such differences must have been caused by movements of the earth's crust.

IMPLICATIONS REGARDING ISOSTASY

The crustal warping recorded by the Chattanooga shale and the Dakota sandstone raises many interesting problems. What forces have raised some parts of the Mississippi Valley and depressed other parts? Have these movements disturbed the isostatic condition of the area? Or have they resulted from isostatic readjustments of the crust to differences in load brought about by erosion in one area and deposition in another?

Have the "unloading" of one area by erosion and the "loading" by newly deposited sediments in another led to the isostatic rise of the first and the sinking of the second, with the concurrent slow flowage of the dense subcrustal material beneath from the loaded to the unloaded area? Considerations of scale and size seem to demand that changes of level of the kind recorded in the warped surface of the

Chattanooga shale must be compensated for in some way by transfer of material at depth to maintain or restore isostatic balanced.

SHOWING WARPS ON MAPS

How can we represent broad warps on maps and sections? If the topography is rough, ordinary geologic maps and cross sections. However, in areas of gently dipping beds and low relief, few beds will crop out at the surface, and hence geologic maps are ill adapted to show the details.

If we are engaged in mining a coal bed by open-pit methods (stripping away the overburden above the coal) a dip of only a few feet per mile may be critical to our operation. Suppose at one place 25 feet of overburden must be stripped off to get to the coal bed. If the ground is flat, and the coal bed dips 100 feet per mile (a little more than 1°), it is obvious that half a mile down-dip our overburden will have increased to 75 feet.

Up-dip, in a little more than one-fourth mile, the coal will rise to the surface and be eroded away. If the ground surface is hilly, as in the Wasatch Plateau coal field of central Utah, or if the coal bed is irregularly warped instead of dipping uniformly, it may seem difficult to compute just what areas of coal can be mined economically by open pits and what portions could better be mined from underground tunnels and shafts. To help in the solution of such a problem, we may need to construct a structure contour map of the coal bed.

Structure Contour Maps

A *structure contour* map, like a topographic map, uses contours to represent lines of equal elevation. But, whereas a topographic map depicts the surface of the ground, a structure contour map shows the surface of a single bed as it would appear if all the beds above were stripped away. If we removed all the overburden from a coal bed and then made a topographic map of the newly exposed surface of the coal, the result would be a structure contour map of the coal.

The outcrop of the Hiawatha coal bed is shown by a heavy black line. Notice that the contours showing the ground surface do not quite parallel the line marking the outcrop of the coal bed. For example, there are seven places in the ravines north of Miller Creek where the 8,250-foot contour intersects the coal bed. At each of these points the top of the coal is 8,250 feet above sea level. Thus we have seven control points through which to draw the 8,250-foot structural contour. On the south wall of Left Fork Canyon there are ten points where the coal bed intersects the 8,000-foot topographic contour, giving us ten control points for drawing the 8,000-foot structural contour. Similarly, other control points are obtained by marking the intersections of other topographic contours with the coal. Still other points can be obtained from borings and wells.

For example, a well drilled on the ridge north of the Right Fork of Miller Creek penetrates the top of the coal bed 750 feet below the surface. Since the curb (top) of this well is on the 9,000-foot topographic contour, the coal bed directly below must have an elevation of 8,250 feet. Thus we have another point through which to draw the 8,250-foot structural contour.

The coal bed is overlain by the Blackhawk formation. The thickness of this formation can be measured on the walls of several of the canyons. Five such measurements in different parts of the area give its thickness as 750, 775, 725, 740, and 760 feet. The variation is not systematic in one direction, so we can accept 750 feet as the average thickness.

Therefore, at any point where a topographic contour intersects the top of the Blackhawk formation we can obtain the elevation of the coal bed directly below by subtracting 750 from the elevation of the topographic contour. The position of the 8,000foot structural contour near the west edge of the map was determined in this way: A well located on the 9,750-foot topographic contour at this point does not go deeply enough to penetrate the coal, but at a depth of 200 feet it does cut the top of a thin bed of limestone which, from measurements on canyon walls, is known to lie 550 feet above the base of the Wasatch formation.

The Price River formation lies between the Wasatch formation and the Blackhawk formation and is 250 feet thick, as determined by measurement at several points in the area. Thus at the site of the well the coal bed must lie 200 feet (depth to the limestone) plus 550 feet (distance from top of limestone to base of Wasatch formation), plus 250 feet (thickness of Price River formation), plus 750 feet (thickness of the Blackhawk formation), equals 1,750 feet below the surface.

Since the surface is at 9,750 feet, the coal is at 8,000 feet at this point; we thus have another point through which to draw the 8,000-foot structural contour. Many other elevations on the coal can be determined by these methods, and from them the structure contours on the Hiawatha coal.

Once such a structure contour map has been prepared, it is easy to tell the amount of overburden above the coal. Since the structure contour map represents the surface of the coal bed, and the topographic map that of the ground, the thickness of overburden at any point can be obtained by subtracting the elevation shown on the structure contour map from that shown on the topographic map for the point directly above. Wherever a structural contour and topographic contour cross we can find the thickness of overburden by subtracting the elevation of the structure contour from that of the topographic contour.

Cross Sections with Exaggerated Vertical Scale

For certain purposes, *geologic cross sections drawn with greatly exaggerated vertical scale* are useful in depicting gently dipping rocks. Such exaggeration, of course, distorts the form of the features shown, and gives a wholly erroneous idea of the dip and the size of the structures, but it permits us to plot thin formations that could not otherwise be shown and to emphasize lateral changes in facies of the sedimentary rocks.

Such sections also emphasize, even though they distort, minor changes in dip, small basins, or other minor structures. For example, the structure beneath a portion of the Russian plain. The large cross section is plotted with a vertical exaggeration of 100 to 1. On it the lateral variation of limestone to shale and of salt to various kinds of rock (sedimentary facies) is clearly shown, but such variations could not have been plotted on the section with 10 to 1 exaggeration shown in the same figure, and if the section had been drawn to true scale its entire thickness would have been encompassed within a single pencil line.

But when the scale is exaggerated 100 times some of the beds in the section appear to dip 45 degrees or more, although their true dip is not more than 1 degree.

CONTINENTAL PLATES

Broad warps are the main records of earth movements in the relatively stable parts of the continents. Such stable areas of nearly horizontal sedimentary rocks are called *continental plates*. They differ from mountain belts in their gently warped strata as contrasted to the strongly folded rocks of most mountains. Also, drilling for petroleum shows that the sedimentary rocks of the continental plates are generally only a few hundred, or at most a few thousand, feet thick above a basement of granite and metamorphic rocks, whereas they are generally far thicker in mountain belts.

The great area from the Appalachians to the Rockies (except for the Ouachita, Arbuckle, and Wichita mountains) and extending northward through Canada in a broad belt east of the Canadian Rockies is characterized by this structure. This is the continental plate of North

America. The great plain reaching from eastern Germany to the Urals, then extending from the Urals far into Siberia and southeastward toward Turkestan is another continental plate.

It should be said in passing that broad warps are also found in mountain belts, but there they are less readily recognized because more intense crustal movements largely obscure such relatively gentle deformation as these features record. We pass now to a description of the records of more intense crustal movements. Although folds and faults are not confined to mountain belts, they are as characteristic of those regions as broad warps are of the continental plates.

FOLDS

Folds are the typical features of most mountain chains. They range in size from microscopic crinkles to great arches and troughs fifty miles or more across. Upfolds or arches in rocks are called *anticlines*; downfolds or troughs are *synclines*.

A *monocline* is a flexure in a series of beds which are nearly horizontal on either side of the flexure. Most folds have been greatly modified by erosion. To visualize their forms we often reconstruct some of the eroded beds by drawing dashed lines beyond the truncated ends of the strata. In the monocline and in the anticlines and syncline the eroded upper part of the structures have been partially restored in this way.

Another kind of reconstruction often used to visualize the form of folds. The upper sketch shows a part of the Jura Mountains in Switzerland. The eroded surfaces of the folds are seen on the top of the upper, and a cross section forms the front. One bed of limestone, more resistant to erosion than most of the other beds, forms cliffs and bold ridges at the surface.

The top of this limestone is shown as a heavy black line on the cross section. In the lower sketch the top surface of this resistant stratum is drawn as it would appear if suspended in space, entirely detached from the beds above and below it. From such a reconstruction we can readily visualize the gross form of the folds.

Structure Symbols

One of the best ways to portray a simple fold is by structure contours. If information is too limitec' to get control for structure contours, special *structure symbols* on the geologic map may help bring out the details of the fold.

The block diagram and geologic map of a small fold in the Jackfork formation of Arkans how various kinds of structures may be represented by symbols. On the geologic map the strike and dip of the beds is shown at several places by the *strike and dip symbol*. While the geologist is mapping, he plots these symbols on the topographic map wherever he determines strike and dip.

Note that two layers of shale interbedded with the prevailing sandstone of the Jackfork formation have been crinkled into a series of small, roughly parallel puckers that trend at an angle to the major fold. It is important in mapping to avoid mistaking the strike and dip of the strata in one of these minor puckers from the general strike of the beds in the major fold.

Another feature easily plotted on a geologic map is the position of the *axis* of the anticline. This is the line of opposed dips or, in other words, the line along which each bed exposed at the surface reaches its highest point as it arches over the top of the fold. On a geologic map an anticlinal axis is shown in symbol by a line crossed by opposed arrows. The axis of a syncline is indicated by a line with converging arrows.

Another measurable feature is the inclination of the top of the fold, or its plunge. Note that at the crest of the anticline warped bed plunges eastward into the ground, following the trend of the fold. The measured inclination to the east, determined in de grees from the horizontal, is 8 degrees; this is the *amount of the plunge*. Plunge is indicated on the geologic map by arrowheads and figures showing the direction and amount of the plunge.

Kinds of Folds

After a *plunging fold* is eroded, the different strata that compose it show a curving or arcuate pattern at the surface of the ground. This relation is beautifully displayed in many airplane views of deeply eroded folds. If the fold has no plunge (*upright fold*), and the topography is nearly flat, the beds will crop out as roughly parallel bands on either side of the axis.

Resistant beds such as sandstone and limestone have been etched into ridges and reveal strikingly the details of the structure. Many folds, such as the Paintrock anticline, are *asymmetrical*—the strata on one side of the axis dip more steeply than on the other side. Some are even overturned or recumbent. Superlative examples of overturned and recumbent folds are found in the Alps.

MOUNTAIN BELTS

Study of the figures shows that some of the folds illustrated could perhaps have been formed by uplift and depression much like the broad warps. But some, like the folds seem to demand that the rocks were squeezed together laterally.

Although some folds are isolated, the vast majority are arranged in long linear belts, forming the characteristic structural pattern of the mountain chains of the earth. Many of these folded belts comprise a whole series of roughly parallel anticlines and synclines so arranged as to imply horizontal, rather than vertical, forces as their cause.

For example, many folds are overturned or even recumbent, forms that could hardly be produced without strong lateral mavement. The evidence of such horizontal compression is further strengthened by the association of many such belts with great thrust faults along which the rocks above the fault plane have been shoved over those beneath for distances measured in miles. Detailed evidence of crustal shortening; here, we merely point out that such structures characterize the Appalachians, the Andes, the Alps, and many other mountain chains.

This shortening implies the crowding together of a wide belt of rocks into a much narrower zone. The Appalachians or the Alps occupy such large areas that considerations of strength show that the folds cannot have grown merely by rising into the air above a solid basement like the folds in a rug pushed together over a smooth floor. The crust could not support such a thickened mass over so large an area. As the rocks are crowded together they must sink into the substratum enough to sustain themselves by flotation.

This is proved, by the fact that the plumb bob is not deflected toward the Himalayas as much as it should be if the range were a pile of rock heaped onto an otherwise uniform crust. In other words, during the crustal shortening, the rocks of the folded belt must have been squeezed not only *upward* into the air, but also *downward* into the plastic substratum below, forming a *mountain root* large enough to buoy up the mountain chain. The existence of such a root beneath the Alps, Appalachians, and other ranges has been proved both by studies of gravity and by the records of earthquake waves that have passed through them.

Metamorphism in Mountain Roots

Where deep erosion has revealed the cores of large mountain chains it is common to find that the folded and faulted rocks along the edges of the mountain belt grade gradually into metamorphic rocks in the more deeply eroded interior. Furthermore, many mountain ranges show evidence of having been eroded to almost featureless plains, and then "rejuvenated" by the warping or faulting upward of the mountain stumps to form a new high mountain range. In such "second cycle" mountains the proportion of metamorphic rocks increases. Also, large intrusions of granite or other plutonic rocks appear in increasing volume.

These relations indicate that in the formation of a large mountain belt by crustal shortening the brittle and cold rocks near the surface are broken by faults and thrown into large folds, but deeper within the mountain belt, where the load of overlying rock is greater and the temperature higher, deformation is accomplished mainly by granulation and recrystallization of the mineral grains to form new metamorphic rocks.

Indeed, it seems reasonable that in the deeper parts of a mountain root the temperature may rise sufficiently to melt parts of the metamorphosed root into new masses of magma. It is not surprising, therefore, that the deeply eroded cores of mountain ranges are composed typically of metamorphic and plutonic rocks.

Joints and Faults

Fractures in rocks are classified by geologists as joints or faults. By definition, *joints* are fractures that have merely opened, without differential slip parallel to the walls; whereas *faults* are fractures along which there has been movement. In actual practice this distinction is often difficult; more commonly the separation is based on the scale of the mapping. A fracture cutting a plagioclase crystal, when seen under the microscope, may show a minute displacement of the plagioclase striae and so can correctly be described as a *microfault*.

The same fracture probably would not be seen in a rock outcrop, but if it were observed it would surely be called a joint, for the small offset of the plagioclase striae would go unnoticed. In geologic mapping, fractures showing visible displacement of only a few feet are generally ignored or are mapped as joints; the mapped faults are larger features with considerable displacement.

Practically every rock outcrop shows numerous joints. Because of their profusion and irregularity they are generally omitted from geologic maps. If there is some economic or other reason for showing them (for example, valuable ore veins may be developed along joints), the most common method is to plot their strike and dip by the symbol.Joints and small faults may be formed in several ways, but large faults that can be traced for miles and that displace their walls hundreds or thousands of feet have certainly been formed by crustal movements.

A few times, after an earthquake, the actual direction of movement with respect to sea level has been measured. More generally, the study of the geometric relations of the rocks on either side of the fault shows only the *apparent* movement. Abrupt bending of the strata as they approach the fault (called *drag*) may show us the direction of relative displacement, or we may be able to tell by the *offset* of the beds at the fault plane. Both drag and offset are well shown on several of the normal faults.

From offsets, measured either on the ground surface (horizontal plane) or on canyon walls and cross sections (vertical plane), we can at times determine the actual amount, as well as the direction, of relative movement along a fault. By matching up the position of the Strontian and Foyers granite complex, the Moine thrust fault, and the areas of schists and other kinds of rocks on the two sides of the Great Glen fault of northern Scotland, W. Q. Kennedy, a Scotch geologist, has inferred that a horizontal displacement of about 65 miles occurred on this great fault.

In areas of uniformly tilted strata it is not always possible to tell from the offset whether vertical or horizontal movement occurred along a fault. Either kind of movement can, after erosion, produce the same *apparent* displacement. Consider the examples sketched Block A shows a series of tilted strata cut by a normal fault. B shows the same faulted terrain after erosion has removed the elevated block. Note the offset of the beds at the fault surface. C shows a similar series of beds displaced by a strike-slip fault. Compare B with C and note that the same offset can be produced either by normal faulting followed by erosion or by strike-slip movement. From mere examination of the offset of the beds in block B we would be unable to tell which kind of faulting, or whether a combination of the two, had produced the final result.

If two or more kinds of rock masses with different dips intersect, we can definitely tell the direction and amount of relative motion. A simple example is illustrated in blocks D, E, and F. Here there is a clear difference between blocks E and F. After elevation of part of block D by normal faulting, erosion has caused outcrops of both dike and strata to migrate down the dip, producing offsets in opposite directions. With strike-slip movement (block F), all offsets are in the same direction, despite the difference in dip of beds and dike. Examples like those are rare; in most cases a dike or other intersecting structure is absent, and we cannot tell the direction of relative motion. The word "apparently" in the definition calls attention to this ambiguity.

Folds and faults are related structures and may grade into one another. Some normal faults die out into monoclines along their strike. Some thrust faults are merely broken anticlines, although others show no evidence of having grown from folds. Even strike-slip faults may show interesting relations to folds. The abrupt bending of the fold axes as they approach the faults, particularly near the points where the faults are dying out, indicates that both folds and faults were probably produced during the same period of deformation. The bending occurs over a distance of a few miles and so is not likely to be simply drag.

Although there are many exceptions, most normal faults dip steeply-commonly 65 degrees to vertical. All gradations occur, but most thrusts can be grouped into either *high-angle thrusts* or *low-angle thrusts*. The separation is based on whether the dip is greater or less than 45 degrees, but lowangle thrusts generally have dips of less than 30 degrees and high-angle thrusts greater than 60 degrees. Low-angle thrusts are conspicuous in many mountain zones. Many have displacements measured in miles and attest to great horizontal shortening in the visible part of the crust. Some strike-slip faults also show displacements of many miles. Most strikeslip faults dip steeply and many change direction of dip when traced along the strike. Many normal faults and thrust faults show a little strike-slip movement; we use the term "strike-slip fault" only when this kind of movement becomes predominant.

The whole classification of faults, however, is artificial; not only are there all transitions between strike-slip faults and other kinds, but many highangle thrusts, when traced along their strike, steepen in dip, eventually become vertical, and then dip in the opposite direction. Thus a thrust fault may change along the strike into a normal fault; a well-known example is the Uinta fault along the north side of the Uinta Mountains in Utah. Fault classification is also ambiguous if the fault planes have been later folded and warped.

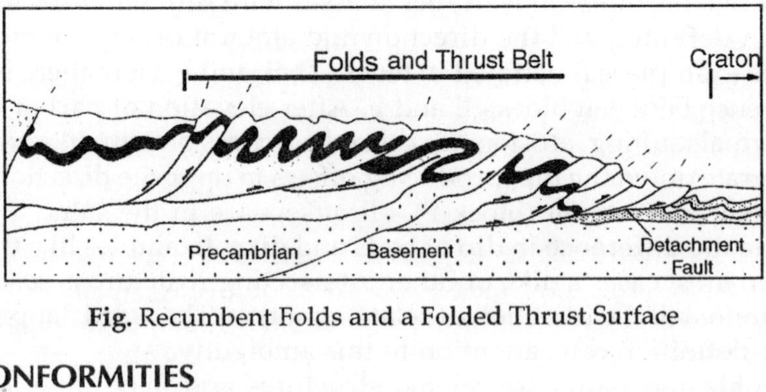

Fig. Recumbent Folds and a Folded Thrust Surface

UNCONFORMITIES

An *unconformity* is a buried erosion surface. To form an unconformity requires a reversal of the conditions of erosion and sedimentation; an area once being eroded has become one of sedimentation. The surface separating the newly deposited rocks from their basement of older rocks is the unconformity.

Typical examples of unconformities well-stratified sandstones, forming the cliff, rest on granite. The surface of contact between the sandstone and granite is the unconformity. Examination of this contact shows clearly that the granite did not invade the sandstone, for there is no metamorphism of the sandstone by heat, nor does the granite penetrate the sandstone in dikes.

On the contrary, the basal layers of the sandstone contain pebbles of the granite, and quartz and feldspar grains derived from the granite abound in it, proving that the sandstone is younger one-third of the way down from the top of the hill. The tilted beds of sandstone below it end abruptly against the surface of the unconformity. Above the unconformity and parallel with it lie beds of much-less-consolidated sandstone and clay.

Their basal layers contain fragments of the underlying tilted rocks. Obviously, the younger poorly consolidated sediments rest on an old erosion surface that had been developed across the older sediments long after they had been laid down, consolidated, and tilted by movements of the earth's crust.

From these two examples it is clear that widespread unconformities record at least three important geologic events:

(1) formation of the rocks below the unconformity,

(2) erosion of these rocks, usually to a comparatively flat surface,

(3) burial of this surface beneath younger strata.

Most unconformities also record crustal movement between (1) and (2) giving the elevation necessary for erosion, but under some conditions—the erosion of a volcanic cone, for example—no crustal movement is required. Many unconformities also record a second movement of the crust between (2) and (3) during which the eroded surface was warped below sea level in order to receive sediments; but, again, a land area might be buried by river-borne sediment without movement of the immediate area of the unconformity.

The burial of an unconformity beneath new sediments is slow. Within the lifetime of one man no significant changes may be visible, yet, even today. slow implacable crustal movements whereby broad erosion surfaces are warped beneath the sea to become buried under sediments are in progress at many places. The drowned topography of the Sunda Shelf and of the subsidence in Denmark, Japan, the northeastern United States, and elsewhere. In each of these regions marine sediments are slowly burying the drowned surface, and ultimately, if not interrupted by a new uplift, they will cover it completely.

Yet, though the erosion surface may disappear as a topographic feature, it remains as an unconformity and thus permanently records a definite series of events in the history of the earth. Unconformities, and the historical record they reveal, are not easily destroyed. Many are recognizable even after the rocks have been folded and metamorphosed. Only widespread uplift and subsequent deep erosion can obliterate them completely.

Many unconformities are covered by continental instead of marine deposits. Every river forms an unconformity when it sidecuts its banks and deposits a layer of gravel over the cut surface. Locally, river deposits bury extensive areas. In the Chaco region of Paraguay and in western Brazil the steep tributaries of the Parana and Amazon rivers are bringing vast quantities of detritus from the Andes and strewing it over low-lying plains at their foot.

These sediments are slowly burying an erosion surface earlier developed across both the sedimentary rocks of the Andean foothills and the plutonic and metamorphic rocks of the Brazilian shield.

Basal Conglomerate

In general, when a broad erosion surface of low relief is submerged it is at a rate so slow that the ocean shore migrates inland perhaps no more than a few yards per century. Hence the sediments that rest upon the unconformity at one point may differ in age from those which cover it a few miles away, even though the unconformity extends continuously between.

The slow advance of the sea upon an erosion surface produces several characteristic effects. Nearly every erosion surface has a mantle of soil and disintegrated rock and shows at least a few rounded hills and other irregularities. Waves are powerful agents of erosion. As the sea slowly advances upon the land the waves have ample time to strip away the soil and perhaps some of the bedrock beneath.

Irregularities on the erosion surface are planed down by wave attack, finally producing a nearly flat bedrock surface strewn with storm-swept gravel and sand. Finer debris is sorted out by the waves and carried into deeper water. The thin layer of conglomerate and coarse sandstone directly overlying the planed-off rock floor is called a *basal conglomerate*. Such conglomerates are the nearshore deposits of the slowly advancing sea.

In a few places submergence has been rapid enough to bury and preserve the soil and minor irregularities of the former land surface.

Most such areas were probably inland arms of the sea, protected from the strong wave at tack of the open ocean. The drowned drainage features on the floor of the southern end of San Francisco Bay.

Twenty-five miles to the west are gravel and coarse sandy beaches of the open Pacific. The same submergence that lowered the floor of San Francisco Bay also affected the coastal section, but here strong wave attack along the open ocean has obliterated all evidence of subaerial topography. All the soil has been removed and a wave-cut platform covered with coarse gravel and sand is developing across the bedrock.

Unconformities covered by continental deposits may show quite different relations. Soils and weathered bedrock are not rare beneath stream deposits. Erosion surfaces buried under lava flows or tuffs may preserve every detail of the former landscape, including such ephemeral features as tree stumps still standing rooted in the position of growth.

KINDS OF UNCONFORMITIES

In cross sections of six simple unconformities from different areas are sketched. Note in A and C that the beds both above and below the surface of unconformity are parallel. Such unconformities are called *disconformities*. Many disconformities are difficult to recognize; the series of beds may appear to be *conformable*, that is, to have been deposited continuously without erosional breaks.

In C, however, disconformity is certain because Silurian strata are absent, as proved by fossils collected from the beds immediately above and below the unconformity. Obviously, the area could not have been under the sea throughout all Silurian time without receiving some deposits. In A the break recorded by the unconformity is vastly greater, embracing nearly all of geologic time since the Cambrian.

Unconformities in which the beds above the unconformity transgress the eroded edges of folded and tilted beds, are called *angular unconformities*. In another common kind of unconformity bedded rocks rest on an eroded surface of plutonic or metamorphic rocks such as granite, gneiss, or gabbro. The term *nonconformity* has been used for this kind of unconformity, but, unfortunately, this term is also used in some geologic literature to describe angular unconformities.

Dating Geologic Events by Unconformities

At Siccar Point, Scotland, the Ordovician and Silurian beds were obviously folded before the unconformity was formed, because the overlying Devonian sandstone is unaffected by the folds. On the other hand, the beds in central Alabama were folded after the development of the unconformity, as both sets of beds as well as the unconformity itself have been arched by the folds.

The relations show merely that the folding was later than the deposition of the Devonian shale, and may have occurred at any time between Devonian and the present. At Siccar Point, however, the folding must have occurred after the deposition of the Ordovician and Silurian strata but before the Devonian sandstone had accumulated. The unconformity shown in D definitely proves that the tilted lower Pleistocene beds of southern California were folded during the Pleistocene because undeformed Pleistocene beds lie above the unconformity.

In the unconformity "Algonkian" of the Grand Canyon was invaded by a dolerite sill and cut by a normal fault before Upper Cambrian time, because the basal nonconformity of the Upper Cambrian truncates these features as well as the tilted "Algonkian" beds and the granite upon which the "Algonkian" itself rests noncomformably.

Similarly, the relations of the two different unconformities shown in F-one beneath the Eocene beds, the other at the base of the Miocene lava flow —tell us that the fault cutting these unconformities has undergone two periods of movement—one during post-Eocene but pre-Miocene time, the other after the faulted area had been eroded to a plain and the Miocene lava spread out over it.

Time Significance of Unconformities

Every unconformity marks a time interval not represented by sedimentary deposits at that particular spot. Either there were no strata deposited here during this time interval because the land was being eroded, or, if any deposits were laid down early in the time interval, they have been eroded away during the later part. It is important to try to estimate the time value of an unconformity.

A common fallacy is to assume that an angular unconformity necessarily implies a greater time lapse than a disconformity. True, in some areas the unconformities that are most widespread and that show the greatest time lapse happen to be angular unconformities, but this is not true in general. Contrast, for example, the unconformities. At several places in the Baltic region clays deposited in Pleistocene lakes lie directly on marine clays that contain Cambrian fossils. The interval of time represented by the disconformity includes all of the Mesozoic and nearly all of the Paleozoic and Cenozoic—a gap of about 400 million years.

Yet, in some places the unconformity can hardly be located, so similar are the Cambrian clays to those of the Pleistocene. In, however, the marine sediments both above and below the angular unconformity are of Pleistocene age, by their abundant fossils. Thus the events during Pleistocene time in southern California included deposition of the lower Pleistocene strata; folding, upturning, and erosion of the newly deposited beds to a comparatively flat surface; sinking of the erosion surface below the sea; and deposition of the upper Pleistocene strata. All of these local events have taken place in perhaps not more than one million years.

Such examples make it clear that the angle between the beds above and below an unconformity is independent of the time represented, and depends only on the nature of the deformation of the older beds. The time interval can only be established by the fossil content of the adjacent rocks.

One reasonably safe generalization, however, is that a long lapse of time—per haps equivalent to at least a geologic epoch—is recorded by a nonconformity where sedimentary rocks lie above coarse-grained plutonic and metamorphic rocks. Since such rocks are formed at considerable depths within the crust, to expose them at the surface must have required a long period of erosion.

DOWNSLOPE MOVEMENTS OF SOIL AND ROCK

Varieties of Downslope Movements

The masses of rock and soil moved directly downslope by the force of gravity range in size from tiny bits of rock rolling down fissured cliffs, or small masses of water-soaked soil gliding down a hillside, to great landslides thousands of feet across crashing down the slopes of mountain ranges.

Some downslope movements are slow distortions (*creep*) of soil, rock, or other incompletely consolidated materials. Typically, they take place on grass-covered slopes where there may be no obvious surface indications of movement.

A second group of *slow slides* and *slow debris flows* differs from creep by showing at the surface distinct boundaries of the individual slides or flows. Such a slide moves as a unit bounded by a well-marked slip surface. The fragments of a debris flow are jumbled together during movement. Slides and flows rarely cover whole hill slopes, but are localized into tongues or broad rounded masses. They leave gaping cracks or rough scars at their sources. Slow slides and debris flows range from a few feet to several miles in length.

A third group, characterized by rapid movement, comprises *rapid debris flows, rapid landslides,* and *rock falls*. Rock falls include small blocks loosened from steep rock surfaces and rattling down in brief free falls to join the accumulations of similar blocks called talus piles. Most of the rapidly moving masses are small, but some great rock falls and landslides have destroyed villages and dammed rivers.

Rainwash

Rainwash on hill slopes marks the initial stage in the movement of the surface waters that are gathered later into well-defined streams and rivers. It is considered here because its effects cannot always be distinguished from movements of water-soaked soil, or even from the downslope creep under gravity of relatively dry masses of soil and rock.

During a heavy rainstorm, hill slopes may be covered by a thin sheet of moving water heavily charged with mud, silt, and humus, churned up by the pelting rain drops. Some of the water soaks into the ground; some moves down the slope.

Generally, such sheets of flowing water quickly separate into numerous small, shallow rills, coursing almost straight downslope, but often dividing, reuniting, and finally losing their individuality at the edge of a stream or on the nearly flat valley floor. Most of these shallow rill-courses are short-lived furrows that may be filled and obliterated in a single season, but some grow into larger, steep-walled permanent trenches called *gullies*.

Creep

Creep is the most widespread of all downslope movements. Typically, it is a matter of slow downhill distortion in soil and mantle rock, or in thinbedded or poorly consolidated sediments. Ordinarily, the movement dies out gradually at shallow depth without developing a slip surface at its lower boundary. Distortion caused by creep is shown by many rock bodies whose previous shape or structure is known.

For example, decayed boulders that once must have been nearly round are now streaked out into thin ribbons of varicolored soil. Creep is also well displayed in the bending of thin-bedded, steeply inclined strata as they come to the surface on a steep hillside.

On many slopes, creep is perceptible through its effects on rock structures, trees, posts, buildings, etc.. Observations extending over several years in climates with marked wet and dry seasons show that soil creeps most actively during the wet season and at the beginning of the dry season.

Fig. Common Effects of Creep

The *mechanisms of creep* are somewhat conjectural. Wetting of clay-rich soil—or freezing of water contained in pores in the soil—causes the soil to swell upward. A particle of soil at *a* is pushed upward at right angles to the surface by the swelling, and comes to rest at point *b*. On drying of the clay, or melting of the ice, the soil contracts again, but the particle will not return to point *a*, but will move vertically to c, or, if the soil is wet enough to flow slightly under the action of gravity, it may glide still farther down the hillside.

By repeated wetting, or freezing, the particle is moved, step by step, farther and farther downslope. Because ice expands 9 per cent on freezing, the upward swelling may be very marked in regions with cold winters. It is called *frost heaving*. The thickening of ice layers in cracks after repeated freezing adds to the heaving. Soil crusts may be raised several inches and paved roads may be heaved upward and broken into rubble by the expansive force. Upon thawing, the shrinkage does not follow exactly the original direction of expansion because the pull of gravity causes a net movement downslope.

Other processes commonly involved, varying with the climatic conditions, include wedging by plant roots; the moving of soil by earthworms, rodents, and other burrowing animals; and the downslope movement of the soil beneath the feet of animals such as sheep or cattle. Any random movement of loose material gives gravity a chance to make a net displacement downslope.

Creep passes imperceptibly into downslope movement of more definitely bounded masses. In the California Coast Ranges a black sandy clay soil develops on several types of sedimentary rock. On many moderate slopes (e.g., 15°-20°) this soil and broken rock are intimately mixed but are sharply separated from the fresh rock beneath by a smooth surface, suggesting that the surficial layer in slow downslope transit across the underlying rock has scraped it clean.

Marked movement in the last 40 or 50 years is shown by the tilting of fence posts and telephone poles, but the lack of scars or irregularities on the surface indicates that the movement is general rather than concentrated in landslide tongues.

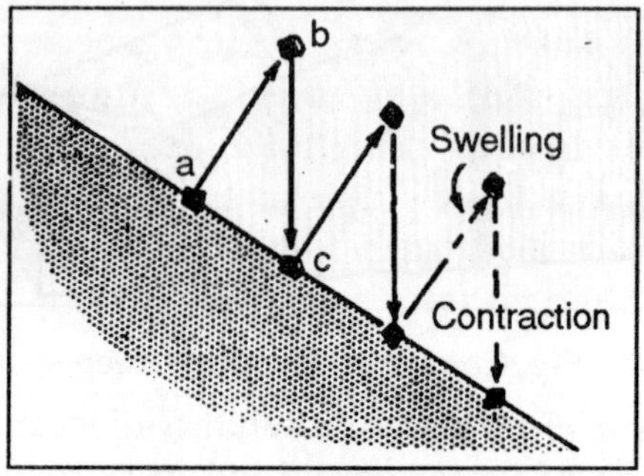

Fig. Downslope Creep

Solifluction

In arctic and subarctic regions, mass movements called *solifluction* (literally: "soil flow") are intermediate between creep and debris flows. Here, frost action and other forms of weathering produce abundant rock fragments of all sizes, including some material fine enough to be called soil. When saturated with water, these accumulations of debris may move slowly down moderate slopes (5°-20°) and along the floors of steep valleys in the form of sheets, lobes, and tongues.

A. Lincoln Washburn, a student of the Arctic, has suggested, from studies on Victoria Island northwest of Hudson Bay, that most of the movement takes place immediately after the thawing of the frost-heaved surface in spring. Rows of stakes Washburn had driven in the lobe indicated slight local tilting and forward movements during the summer of 1940.

The maximum forward movement of 1¾ inches occurred almost entirely in the 33 days beginning June 14. Apparently in arctic and subarctic regions, solifluction proceeds by very slow flow of water-saturated soil, year after year. Solifluction is also prominent above timber line on mountains in temperate regions.

ROCK GLACIERS

In southwestern Colorado and elsewhere, the flat floors of large amphitheaters are strewn with fine debris marked by surficial ridges somewhat like those on solifluction lobes, but also resembling those that form on rock falls. These ridged debris piles have been called *rock glaciers*, perhaps because the ridges resemble the debris ridges on glaciers. Rock glaciers probably do not result from rock falls or rapid slides, because the debris is relatively fine grained and the ridges are sinuous and irregular, but the mechanisms of their motion are somewhat uncertain.

Slow Slides and Flows

If part of the soil and loose rock on a hillside moves downslope a little more rapidly than the general hillside creep, it is set off from its surroundings by a gaping crack at its upper edge, and by cracks or ridges at its sides. If such a mass glides more or less as a unit along a sharply defined basal slip surface, it is a slide; if its constituent parts are broken up and jumbled together during movement, it is a *debris flow*.

Many such masses are debris flows throughout, but others start as slides and become broken and mixed accumulations farther downslope. Some slow slides and flows are very large. An example from Wyoming will serve to show the distinctive features of the group.

GROS VENTRE DEBRIS FLOW

The Gros Ventre River (pronounced Grow Vont) is a minor westward-flowing tributary of the Snake River south of Yellowstone Park, in Wyoming. In the spring of 1909, the Gros Ventre River was dammed in midcourse by rock debris that had begun to flow as early as May, 1908, and to move slowly down the moderate (10°-20°) slope of the Gros Ventre Mountains on the south side of the river. The rocks involved in the sliding were soft shales, with some interbedded thin sandstone and limestone layers.

The shales had been thoroughly soaked by heavy rains. Some of the strata probably slipped along the bedding planes, which almost parallel the land surface. The sliding was imperceptible to an observer but wrought notable changes in the landscape within a few weeks. Telephone poles tilted slowly downhill, snapping the wires. A wagon road paralleling the river soon became so hopelessly twisted and broken that repairs were futile. Eventually, it was so obliterated by movement that traces of it were hard to find.

The flow did not move in one mass, but in sections, beginning at the east side and spreading week by week. The sliding and crumbling mass moved fastest in the wet spring months of 1909 and slowed noticeably by autumn. Nevertheless, incessant movement toward and into the river bed continued through 1910. It almost entirely ceased in 1911; the river was then able to cut 10 feet down into the debris dam and partially drain the lake that had formed on the upstream side.

Numerous gaping fissures formed at and near the places of initial movement on the east and south sides of the Gros Ventre debris flow. Further down, where the material of an earlier flow, faintly indicated, was set in renewed motion, earthen domes swelled up and broke open in broad cracks.

The mass as a whole thickened toward the river, developed a very irregular surface, and was churned into a jumbled mixture of clay and coarser material.

Slow Slides

Many slides break up far less completely than did the Gros Ventre flow. A mass of rock along a cliffed shore may move on a flat, water-lubricated slip surface for many feet without breaking up at all, leaving only a gaping fissure on the inland side as evidence of movement. In weaker materials, small slide blocks commonly rotate on cylindrical slip surfaces (soles) forming a depression at the head of each slide unit.

Fig. Rotation of Slide Blocks on Curved Soles.

RAPID SLIDES., FLOWS, AND FALLS

4Small Rapid Slides

The great majority of rapid gravity movements involve only small volumes of soil or rock, but the aggregate effects are large. In humid, temperate regions, small rapid slides with barely recognizable slip surfaces are extremely common in soil or weak sedimentary rock, especially on steep grassy slopes and after unusually heavy rains.

The depression, or scar, left by a typical slip may be 10 feet wide and 20 feet long in the direction of slope. At the base of the scar, the soil piles up in more-or-less crumpled and disordered masses. The distinctive features of such slides are present in the miniature examples in dry volcanic ash. One of these little slides is very fresh; the others have lost their original sharpness.

TALUS PILES

Quantitatively, the most important falls of rock are countless multitudes of small fragments, from a fraction of an inch to a few feet in diameter, that drop from cliffs or steep rock surfaces and accumulate in talus piles at the bases of the steep slope or cliff. A talus pile maintains a fairly constant surface slope as it grows. The slope angle, commonly about 30°, is called *the angle of repose* because it is the maximum angle at which the material will remain stable.

In their quantitative importance, small rapid slides and talus piles are perhaps second only to creep among gravity movements. The widening of steep-walled valleys in arid regions is largely by the fall and continued streamward movement of large and small talus fragments.

Rate of Talus Formation

Those who climb steep mountain slopes of closely jointed rocks know well the sound of, and also the danger from, falling rock fragments that come from high aloft and bound erratically down long talus slopes. Some talus piles grow rapidly, others slowly.

The mere presence of talus is no proof of rapid change in the configuration of the cliffs above. Great talus piles at the foot of granite cliffs in southern Arizona are composed of huge blocks so thoroughly weathered that they could not have withstood the impact of fall. They must have weathered in place, on the talus pile. Weathering is so slow in this climate that a long time is implied for the talus growth.

In a few places, the rate of spalling has been roughly determined. Castle Rock, Kansas, made of soft sedimentary strata. In the semi-arid Kansas climate, it has suffered the loss of perhaps 3 or 4 per cent of its volume between 1894 and 1941. At this rate, Castle Rock might be consumed in 1,200 to 1,600 years.

Large Rock Falls and Mudflows

Large slides of rock occasionally avalanche down steep slopes, or even break loose and hurtle through the air. Large and rapidly moving mudflows also occur, and, if well lubricated, may move swiftly over very gentle slopes. Several such rapid slides, flows, and falls have destroyed villages or portions of cities. From the many such catastrophes, we select for description a rock slide and fall in the Swiss Alps, and mudflows in Norway and Switzerland.

Elm Rock Slide and Fall

Perhaps the best known rock fall occurred at Elm, a village in the northeastern Swiss Alps, in 1881. A steep slope 2,000 feet high, the prow of a ridge between two valleys, had been undercut half way up in quarrying for slate. In the course of one and one-half years, a curving fissure 30 feet deep had been formed, extending 1,100 feet above the quarry and outlining the future slip mass.

This fissure was approximately perpendicular to the stratification and foliation of the rocks. In the late summer of 1881, the local runoff from heavy rains poured into the fissure, saturating the shattered rocks. Late one September afternoon, two small slides occurred, starting just above and on either side of the quarry. A few minutes later, the whole mass outlined by the fracture crashed down, filled the quarry, and shot part way across the narrow valley as a free-falling rock avalanche.

On striking the valley floor, it rushed obliquely up the opposite slope to a height of 300 feet, turned, and shot down the valley in a debris stream, destroying houses and everything in its path, and killing 115 people. Thirteen million cubic yards of rock fell an average of about 1,450 feet, and spread as rubble over a third of a square mile to a depth of 30 to 60 feet. A dense cloud of dust covered the scene for several minutes.

All observers agreed that, after the sliding mass had fallen steeply to the quarry floor, it shot forward into free fall. The villagers could see across the valley beneath the cascading torrent of shattered stone to the hillside beyond.

The Elm debris was composed largely of fragments a few inches in diameter, including much intermingled soil and rock dust. Many larger blocks, 5 to 20 feet across, were scattered through it, especially in the central, higher, and later portions of the debris tongue. The margins were abrupt. The top of the debris stream was hummocky and irregular, and low ridges festooned its surface. By 1928, 47 years after the event, almost all the devastated area had been restored to pasture and potatoes.

From eyewitnesses' reports, it was estimated that the debris at the front of the tongue traveled one and one-half miles in 45 or 50 seconds. Physical considerations make this appear reasonable. The physical calculations are based on the acceleration of gravity. In free fall the acceleration or increase of velocity is 32 feet per second in each second. The short preliminary sliding and the free fall at Elm were calculated to have taken 17.2 seconds, and the velocity when it hit the valley floor about 186 miles per hour. Including the time of braking to a stop down valley, the whole duration was calculated at 53.4 seconds—in good agreement with estimates by eyewitnesses. The average velocity of the farthest traveled blocks was calculated to have been 93 miles per hour.

A rock fall such as that at Elm illustrates the physical concepts developed concerning the principle of the *conservation of energy*. Before the fall, the poised rock had energy of position, or *potential energy*, measured by its mass and its elevation above a convenient reference plane. During fall, it lost potential energy and gained energy of motion, measured by mass and velocity. When brought to rest by friction, *energy of motion* was changed to *heat energy*.

The law of conservation of energy requires that the sum of the potential energy, energy of motion, and heat energy resident in the rock mass or developed from it remains constant. Hence, the heat energy developed by braking must have been equal to the potential energy lost during the fall. Geologists are necessarily concerned with the source of energy in all geological processes. In dealing with erosion and downslope movements, the principle source is that of the earth's gravitational field, as we have just seen.

Rapid Mudflows

The Elm fall affected hard rocks, and was made possible by a steep slope. Masses of mud, on the other hand, may break through weak barriers and flow fairly rapidly for long distances even on gentle slopes.

In northwestern Europe and eastern Canada, many valleys are floored with unconsolidated clays and sands, extending many miles inland and rising to elevations of several hundred feet. These soft sediments become saturated with water after the spring thaw. A mass of mud as much as a square mile in area may suddenly break through a slightly more solid barrier, and roll down valley as a destructive mudflow, the front moving as much as 3 or 4 miles in a single hour, and then piling up in a temporary dam of sandy clay, behind which a lake forms that in turn overflows and breaches the dam.

In the source area, such a mudflow leaves a clean-cut scar, with rough floor and steep walls a few feet or a few tens of feet high. One night in 1893, a mudflow of this kind moved down a Norwegian valley so fast that 111 persons were caught by it and lost their lives. Others escaped after being carried long distances in their wooden houses.

In arid regions, still another kind of mudflow takes place. A thick mass of silt, sand, and coarser debris scoured by a flood from the canyon floor may roll forward for miles, damming, piling up, breaking through, and finally coming to a stop by gradual thickening through loss of water, or by spreading over a plain.

Some of the largest and most destructive mudflows have developed in the pyroclastic debris ejected from active volcanoes.

UNDERWATER FLOWS

Of all downslope movements of solid or semiliquid material the most obscure and difficult to study are those which glide along the floors of lakes or seas. One thoroughly studied example started at the shore of the Swiss lake of Zug and squirted out under water along the lake floor.

Zug Underwater Flow The source area in the city of Zug extended from a lake front retaining wall for about 200 feet inland to a street lined with buildings. Most of the area was filled land. In the two years after the retaining wall was built, water began to appear in previously dry cellars behind it, and, in the spring of 1887, the ground had settled and the pavements had cracked slightly.

The first disastrous movement, in midafternoon of July 5, dropped a small section of the retaining wall and three houses beneath lake level and caused several deaths. A half hour later, wooden piles from the broken retaining wall suddenly rose to the lake surface a hundred yards offshore, showing that lateral movement as well as sinking had occurred. The main flow came three hours later, with streets and houses suddenly settling beneath the lake level. The average drop was about 25 feet, and some buildings moved 30 to 60 feet lakeward.

The material that flowed out into the lake from beneath the source area was water-saturated silt and fine sand from the submerged part of a small stream delta. The silt flow excavated a trench 200 feet wide, up to 20 feet deep and extending almost 1,000 feet along the lake floor in the similar silt and sand on the bottom of the lake. From the end of the underwater trench a new deposit of sandy silt, from the source area and trench, extended out over the lake floor as a thin debris tongue 2,500 feet long, with an uneven hummocky surface.

The end of the tongue was 140 to 150 feet below the land surface in the source area, so the slope on which it flowed was less than 3°. The deposit was more than 15 feet thick in places, but averaged less. About 200,000 cubic yards of material were moved, somewhat less than half from the land; the remainder eroded from the lake floor by the flow. The Zug flow was a gravity current of silt and sand, heavier than the lake water. It moved down the slope of the lake floor somewhat as a stream of water on land moves downslope beneath a cover of air.

Several other similar debris flows into lakes have occurred in

Switzerland in historic time, including a disastrous earlier flow at Zug, March 4, 1435. All appear to have been caused by the loading of weak sediments by artificial fill and buildings. Because started by man's modification of existing conditions, their geological significance is somewhat doubtful. Nevertheless, there are abundant evidences of similar flows in marine sediments, so we shall refer to them later in connection with submarine topography.

DISTINCTION BETWEEN PRODUCTS OF SLOW AND RAPID MOVEMENT

The division of slides and debris flows into two groups, one whose movement is imperceptible to the casual observer; the other, more rapid, is satisfactory if the movements are actually observed. The great majority of slides and debris flows, however, are recognized only by their effects. In these, the distinction may be difficult or impossible.

Rapid movement usually leaves a well-marked scar or amphitheater in the source area, and commonly a symmetrical pile of rather finely divided debris downslope, often with transverse or concentric corrugations. Slow *slides*, because of their less severe impacts with the valley floors, are less likely to break up into fine fragments, except near their lower margins. Most are composed of units that can be mapped, like fault blocks.

PREHISTORIC SLIDES AND OTHER DOWNSLOPE MOVEMENTS

Thousands of ancient slides have been recognized in all parts of the world. Perhaps the largest is just over the hill from Elm, the enormous Flims landslide. This mass probably started movement as a unit, but has now been thoroughly broken up, especially some portions which had traveled the farthest. It started to move downslope along the slip plane of an Early Tertiary thrust fault.

Long ago, probably in Late Pleistocene time, it blocked the valley of the upper Rhine, forming a lake. Now the Rhine has eroded the landslide dam, cutting a gorge 2,000 feet deep and 9 miles long. The hummocky surface of the great slide, somewhat modified by erosion; the well-marked margins of the landslide dam; and the sudden change in the Rhine from a broad, open upper valley to the steep-walled gorge through the slide, and then to the open valley below. One thing remains uncertain.

A more complex assemblage is found where the San Bernardino Mountains rise above the southern margin of the Mojave Desert in southern Cal fornia. A partly eroded landslide mass is spread out below the outcrops of two thrust faults. On the lower of these faults, granite overrides Late Tertiary sediments; on the upper, crystalline limestone overrides the granite. The major slide block, composed of limestone, also lies on Tertiary sediments; it may have reached its present position in Late Tertiary or Pleistocene time.

The most puzzling element in the assemblage is a broad lobe of poorly cemented breccia that extends 4 or 5 miles out over the flat desert floor and is made up in most places exclusively of pebble-sized limestone blocks. It is somewhat eroded—one stream from the mountains has cut entirely through it—but its margins are still 25 to 100 feet high and sharply marked. Can this be the debris from a giant rock fall, perhaps 100 times as large as that at Elm?

Several features suggest this, particularly the blocks being exclusively of limestone with no admixture of any of the other rocks of the region; the piling up of the breccia against low hills far out on the desert, with a slight deflection around the hills somewhat like that at Elm; and, finally, at the edge of the low hills, the exposure by erosion of layers of two underlying breccias, the upper composed of sheared and crushed granite, the lower of the Tertiary sediments. A somewhat similar but less sharply defined segregation by rock types was present in the Elm debris.

The corrugations of the breccia surface, unlike those at Elm, are almost straight. They nearly parallel the trend of the partially overridden low bedrock hills, and may represent waves set up by these barriers. The differences from the Elm flow, however, raise doubt as to the origin of the limestone breccia, and stimulate study both of this and of any active slides or falls that may in the future be observed.

Slides or Flows Preserved in Sedimentary Rocks

The form and textures of certain breccias, or brecciated strata, now parts of Paleozoic, Mesozoic, and Tertiary formations, suggest that these are ancient slides or debris flows.

Most convincing are distortions in thin-bedded marine sediments and lenses of heterogeneous breccia apparently in troughs like that developed in 1887 on the lake floor near Zug. These have been recognized in Wales, Peru, Quebec, and California, to mention only a few of many localities.

SIGNIFICANCE OF DOWNSLOPE MOVEMENTS

The Grand Canyon of the Colorado River not only gives evidence of extensive erosion, but makes possible at least a rough estimate of the parts played by different erosional processes in forming it. The Grand Canyon approximately from south (left) to north (right), showing the Granite Gorge in which the Colorado River flows, and the stepped profiles in the slightly inclined strata overlying the gneiss of the Granite Gorge.

The total depth of the canyon is about 1 mile, and the width from rim to rim about 7½ miles. The Granite Gorge is about 1,300 feet deep at the line of section, and the inner gorge of Phantom Creek, about 900 feet deep. The Granite Gorge has relatively straight side slopes, the inner gorge of Phantom Creek has side slopes that are concave upward. Phantom Creek furnishes an especially favorable case for analysis of the relative roles of stream transport and downslope movements in denudation.

The stream follows a fault of 100 or 150 feet vertical displacement. The position of the stream was apparently determined by the fault; presumably the stream has eroded downward along the fault, fixed in position by the ready erosion of the shattered rock along the fault zone. The dashed lines indicate a trench as wide as the creek bed to represent the erosion ascribed to the stream. But if the stream directly eroded only this narrow vertical slot, what processes account for the much wider actual valley?

Judging the past in terms of the present, these other processes include rainwash concentrated into minor rills and along better defined gullies, and also several kinds of downslope movements of rock. In this arid climate the initial fall and continual streamward movement of talus fragments is an important cause of the widening of Phantom Creek valley, but many other kinds of downslope movement also contribute.

The Grand Canyon itself has presumably been cut by the Colorado River and widened by rainwash and downslope movements in the same manner as Phantom Creek. Since the course of the Colorado was not determined by a fault line, however, the river may have wandered somewhat to reach its present position instead of cutting straight down, as did Phantom Creek.

A similar development may be taken as probable for streams and stream systems in other regions, especially those in canyons or deep valleys. Going one step further, we may assume tentatively that in a stream-cut valley whose floor is little or no wider than the stream, the depth of the valley is a measure of downcutting by the stream and the valley's width is an indication of the amount of material contributed by rainwash and downslope movements of rock and soil. Steep side slopes indicate the predominance of downcutting by the stream, gentle slopes the effectiveness of the widening process.

Downslope movements of rock and soil are, of course, not limited to the matched sides of valleys. They are equally effective in reducing the steepness of single slopes or cliffs, as along shores. Gravity movements extend to the very summits of ridges, reducing their height.

Rocky summits are lowered by the fall of fragments loosened by weathering. This process has contributed to the reduction of the summit ridge between Phantom Creek and the Grand Canyon, so that it now lies well below the level of the plateaus to north and south. Soil-covered divides, on the other hand, must be lowered chiefly by creep and rainwash.

Effect of Plant Cover

Downslope movement of material toward drainage lines varies with the plant cover, which, in turn, is dependent primarily upon climate. With regard to plant abundance, the most important distinction among nonarctic climates is that between humid and arid climates.

Even in humid regions, the density of the plant cover varies widely, being thin on cold moors and very thick in tropical areas with year-round rainfall. Typically, as in eastern United States or west-central Europe, trees, shrubs, and grass cover almost the whole surface, and topographic forms are smoothly rounded.

Here, creep is the principal variety of downslope movement. Minor differences in rock resistance, as on a single slope, are masked by the almost continuous cover of soil and vegetation, but major differences become apparent in the steepness of the side slopes and in the general pattern of ridges and valleys. The etching out of weaker rocks is called *differential erosion*. In arid regions, trees and shrubs are few, and grass may be absent. As a result, no soil covers the resistant rocks, which stand out in knobs, ridges, or cliffs. Weak rocks are reduced more rapidly by downslope movements and rainwash, and their areas of outcrop are obscured by talus accumulations derived from the resistant rocks.

Such differences in rock resistance through differential erosion become strikingly apparent even when slight, as in the set of horizontal strata. On a larger scale, compound cross profiles of hill slopes resulting from differential erosion are more sharply defined in arid than in humid regions. Take place on a great scale, even though individual pieces of the mass involved are still strong by our human standards. Downslope movement is a very important and widespread process of erosion, but it rarely goes on alone. Streams cut valleys, gravity movements widen them, and the streams carry away the debris. Glaciers gouge valleys deeper, and gravity movements load the glacier margins with rock fragments that the ice transports away. Waves driving on shore undermine cliffs, whose upper portions then fall into the sea, to be broken up and carried away by wave-generated currents. In brief, the role of gravity movements is to supply material to the agents of long-distance transportation.

Downslope movements differ from most other erosional processes in that they can operate below the sea as well as above. Though the bouyant effect of the water lessens the effective pull of gravity on material immersed in it, the soaking of the material is complete. As a result, even though weathering is thought to be slight below the sea, downslope movements should be common there. The record of their existence in the geologic past is found in disturbed beds and brecciated areas in the marine sedimentary rocks.

It is probable that such movements play a large part in molding the submarine topography at depths far below sea level. Thus, downslope movements may contribute to some extent to both of the two broad levels of the earth—one near sea level, the other corresponding to the floor of the ocean. Downslope movements on land bring debris to the streams, and they in turn build the shorelines seaward. Downslope movements in the oceans tend to reduce submarine heights to the level of the ocean floor.

Engineering Applications

Gravity movements of soil and rock may affect man-made structures such as roads, bridges, dams, or houses. In the United States alone, clearing highways of the debris that slides or rolls from the slopes above, and the repair of sections of roadbed that have settled and slumped downslope, costs millions of dollars annually. Many newly built railroads and highways have had to be rerouted within a few years in order to detour slide areas or cliffs which shed many talus fragments. Expensive mistakes of this kind can be avoided by careful geologic inspection of the ground along the right-of-way. Areas of active sliding show readily recognized features: The hill slope is generally hummocky and con tains undrained depressions, the rock may show slip surfaces roughly parallel to the surface of the hill, landslide scars and curving debris ridges are likely to be present, fences and telephone poles are tilted downhill, and tree trunks bend uniformly as they enter the ground.

Obviously, the cut made for a highway or railroad increases the danger of sliding because the excavation removes support on the downslope side of the slide. Many an ancient landslide or debris flow that had not moved for tens or even thousands of years has been reactivated by the removal of debris from its toe during the construction of a roadbed. Although highways and railroads can generally be relocated to detour a dangerous slide area, it is sometimes impossible to move other kinds of structures that may be endangered by gravity movements. Geologists and engineers must then take steps to stop, or at least to slow down and minimize, the danger. An oil field near Ventura, California, affords a typical example. Oil wells drilled through a slide were slowly bent out of line as it moved. Some of the well casings were sheared completely off at a slip surface along the base of the slide.

Movement was most rapid in the winter when the ground was saturated by seasonal rains. Unless the slide could be controlled, a valuable oil field would have to be abandoned. Geologists and engineers solved the problem by digging and boring galleries along the base of the slide and installing drain tile to carry off the seepage from the rains. They also paved the entire hillside with asphalt. Most of the rain now runs rapidly off the pavement, and the little that does penetrate the slide is drained off. By thus preventing access of water, which transformed the clays into a lubricant, they stopped the motion of the slide.

During the building of the huge Grand Coulee Dam on the Columbia River, construction work was threatened when a tremendous mass of watersoaked silt and sand started to creep into the excavation dug for the north abutment of the dam. To stop this threatened slide, engineers hit on the ingenious idea of penetrating the silts with numerous pipes in which a refrigerant was circulated. The refrigerant froze the water in the pores of the silt, thus cementing the particles and increasing the strength of the silt. This effectively stopped the movement, and the area was kept refrigerated until the concrete for the dam was poured, and the excavation thus filled and stabilized. Similar problems must be met when heavy structures such as large bridges and dams are built on soft clay or on creeping ground. Many soils behave plastically when loaded, and will flow radially outward from beneath the load. One of the piers of the San Francisco Bay Bridge was made purposely large at its base so that the weight of the heavy structure could be distrib uted over a wider area of the hard clay on which the pier was set. Many kinds of laboratory tests have been devised to determine the load that various kinds of sand, clay, and other loose foundation materials will bear.

These tests are the basis of a relatively new branch of engineering science called *soil mechanics*. Field conditions, including attitude of stratification or other slip planes, amount of contained water, slope, and other factors must be taken into account, as well as the laboratory characteristics of the materials. By these methods it is often possible to forestall or control gravity movements, even in places where weak materials are covered by massive structures.

Chapter 11

Causes of Earthquake

The sudden movement of the Earth caused by the abrupt release of accumulated strain along a fault in the interior. The released energy passes through the Earth as seismic waves (low-frequency sound waves), which cause the shaking. Seismic waves continue to travel through the Earth after the fault motion has stopped. Recordings of earthquakes, called seismograms, illustrate that such motion is recorded all over the Earth for hours, and even days, after an earthquake.

Earthquakes are not distributed randomly over the globe but tend to occur in narrow, continuous belts of activity. Approximately 90% of all earthquakes occur in these belts, which define the boundaries of the Earth's plates. The plates are in continuous motion with respect to one another at rates on the order of centimeters per year; this plate motion is responsible for most geological activity.

Plate motion occurs because the outer cold, hard skin of the Earth, the lithosphere, overlies a hotter, soft layer known as the asthenosphere. Heat from decay of radioactive minerals in the Earth's interior sets the asthenosphere into thermal convection. This convection has broken the lithosphere into plates which move about in response to the convective motion.

As the plates move past each other, little of the motion at their boundaries occurs by continuous slippage; most of the motion occurs in a series of rapid jerks. Each jerk is an earthquake. This happens because, under the pressure and temperature conditions of the shallow part of the Earth's lithosphere, the frictional sliding of rock exhibits a property known as stick-slip, in which frictional sliding occurs in a series of jerky movements, interspersed with periods of no motion—or sticking. In the geologic time frame, then, the lithospheric plates chatter at their boundaries, and at any one place the time between chatters may be hundreds of years.

The periods between major earthquakes is thus one during which strain slowly builds up near the plate boundary in response to the continuous movement of the plates. The strain is ultimately released by an earthquake when the frictional strength of the plate boundary is exceeded.

Most great earthquakes occur on the boundaries between lithospheric plates and arise directly from the motions between the plates. These may be called plate boundary earthquakes. There are many earthquakes, sometimes of substantial size, that cannot be related so simply to the movements of the plates.

At many plate boundaries, earthquakes occur over a broad zone—often several hundred miles wide—adjacent to the plate boundary. These earthquakes, which may be called plate boundary-related earthquakes, are secondarily caused by the stresses set up at the plate boundary. Some earthquakes also occur, although infrequently, within plates. These earthquakes, which are not related to plate boundaries, are called intraplate earthquakes. The immediate cause of intraplate earthquakes is not understood.

In addition to the tectonic types of earthquakes described above, some earthquakes are directly associated with volcanic activity. These volcanic earthquakes result from the motion of undergound magma that leads to volcanic eruptions.

Earthquakes often occur in well-defined sequences in time. Tectonic earthquakes are often preceded, by a few days to weeks, by several smaller shocks (foreshocks), and are nearly always followed by large numbers of aftershocks. Foreshocks and aftershocks are usually much smaller than the main shock. Volcanic earthquakes often occur in flurries of activity, with no discernible main shock. This type of sequence is called a swarm.

Earthquakes range enormously in size, from tremors in which slippage of a few tenths of an inch occurs on a few feet of fault, to the greatest events, which may involve a rupture many hundreds of miles long, with tens of feet of slip.

The size of an earthquake is given by its moment: average slip times the fault area that slipped times the elastic constant of the Earth. The units of seismic moment are dyne-centimeters. An older measure of earthquake size is magnitude, which is proportional to the logarithm of moment. Magnitude 2.0 is about the smallest tremor that can be felt. Most destructive earthquakes are greater than magnitude 6; the largest shock known was the 1960 Chile earthquake, with a moment of 10^{30} dyne-centimeters (10^{23} newton-meters) or magnitude 9.5. It involved a fault 600 mi (1000 km) long slipping 30 ft (10 m).

The intensity of an earthquake is a measure of the severity of shaking and its attendant damage at a point on the surface of the Earth.

The same earthquake may therefore have different intensities at different places.

The intensity usually decreases away from the epicenter (the point on the surface directly above the onset of the earthquake), but its value depends on many factors and generally increases with moment. Intensity is usually higher in areas with thick alluvial cover or landfill than in areas of shallow soil or bare rock. Poor building construction leads to high intensity ratings because the damage to structures is high. Intensity is therefore more a measure of the earthquake's effect on humans than an innate property of the earthquake.

An earthquake is the result of a sudden release of energy in the Earth's crust that creates seismic waves. Earthquakes are recorded with a seismometer, also known as a seismograph. The moment magnitude of an earthquake is conventionally reported, or the related and mostly obsolete Richter magnitude, with magnitude 3 or lower earthquakes being mostly imperceptible and magnitude 7 causing serious damage over large areas. Intensity of shaking is measured on the modified Mercalli scale.

At the Earth's surface, earthquakes manifest themselves by shaking and sometimes displacing the ground. When a large earthquake epicenter is located offshore, the seabed sometimes suffers sufficient displacement to cause a tsunami. The shaking in earthquakes can also trigger landslides and occasionally volcanic activity.

In its most generic sense, the word *earthquake* is used to describe any seismic event—whether a natural phenomenon or an event caused by humans—that generates seismic waves. Earthquakes are caused mostly by rupture of geological faults, but also by volcanic activity, landslides, mine blasts, and nuclear experiments. An earthquake's point of initial rupture is called its focus or hypocenter. The term epicenter refers to the point at ground level directly above this.

NATURALLY OCCURRING EARTHQUAKES

Tectonic earthquakes will occur anywhere within the earth where there is sufficient stored elastic strain energy to drive fracture propagation along a fault plane. In the case of transform or convergent type plate boundaries, which form the largest fault surfaces on earth, they will move past each other smoothly and aseismically only if there are no irregularities or asperities along the boundary that increase the frictional resistance. Most boundaries do have such asperities and this leads to a form of stick-slip behaviour.

Once the boundary has locked, continued relative motion between the plates leads to increasing stress and therefore, stored strain energy in the volume around the fault surface. This continues until the stress has risen sufficiently to break through the asperity, suddenly allowing sliding over the locked portion of the fault, releasing the stored energy. This energy is released as a combination of radiated elastic strain seismic waves, frictional heating of the fault surface, and cracking of the rock, thus causing an earthquake.

This process of gradual build-up of strain and stress punctuated by occasional sudden earthquake failure is referred to as the Elastic-rebound theory. It is estimated that only 10 per cent or less of an earthquake's total energy is radiated as seismic energy. Most of the earthquake's energy is used to power the earthquake fracture growth or is converted into heat generated by friction. Therefore, earthquakes lower the Earth's available elastic potential energy and raise its temperature, though these changes are negligible compared to the conductive and convective flow of heat out from the Earth's deep interior.

CAUSES OF EARTHQUAKE

An Earthquake is a sudden tremor or movement of the earth's crust, which originates naturally at or below the surface. The word natural is important here, since it excludes shock waves caused by French nuclear tests, man made explosions and landslides caused by building work.

There are two main causes of earthquakes:

- Firstly, they can be linked to explosive volcanic eruptions; they are in fact very common in areas of volcanic activity where they either proceed or accompany eruptions.
- Secondly, they can be triggered by Tectonic activity associated with plate margins and faults. The majority of earthquakes world wide are of this type.

An earthquake can be likened to the effect observed when a stone is thrown into water. After the stone hits the water a series of concentric waves will move outwards from the centre. The same events occur in an earthquake. There is a sudden movement within the crust or mantle, and concentric shock waves move out from that point. Geologists and Geographers call the origin of the earthquake the focus. Since this is often deep below the surface and difficult to map, the location of the earthquake is often referred to as the point on the Earth surface directly above the focus. This point is called the epicentre.

The strength, or magnitude, of the shockwaves determines the extent of the damage caused. Two main scales exist for defining the strength, the Mercalli Scale and the Richter Scale.

Earthquakes are three dimensional events, the waves move outwards from the focus, but can travel in both the horizontal and vertical plains. This produces three different types of waves which have their own distinct characteristics and can only move through certain layers within the Earth. Lets take a look at these three forms of shock waves.

TYPES OF SHOCKWAVES

P-Waves

Primary Waves (P-Waves) are identical in character to sound waves. They are high frequency, short-wavelength, longitudinal waves which can pass through both solids and liquids. The ground is forced to move forwards and backwards as it is compressed and decompressed. This produces relatively small displacements of the ground. P Waves can be reflected and refracted, and under certain circumstances can change into S-Waves.

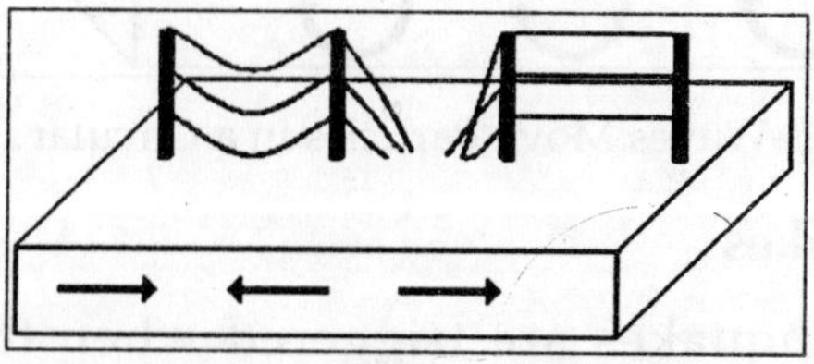

Fig. Particles are Compressed and Expanded in the Wave's Direction.

S-Waves

Secondary Waves (S-Waves) travel more slowly than P-Waves and arrive at any given point *after* the P-Waves. Like P-Waves they are high frequency, short-wavelength waves, but instead of being longitudinal they are transverse. They move in all directions away from their source, at speeds which depend upon the density of the rocks through which they are moving. They cannot move through liquids. On the surface of the Earth, S-Waves are responsible for the sideways displacement of walls and fences, leaving them 'S' shaped.

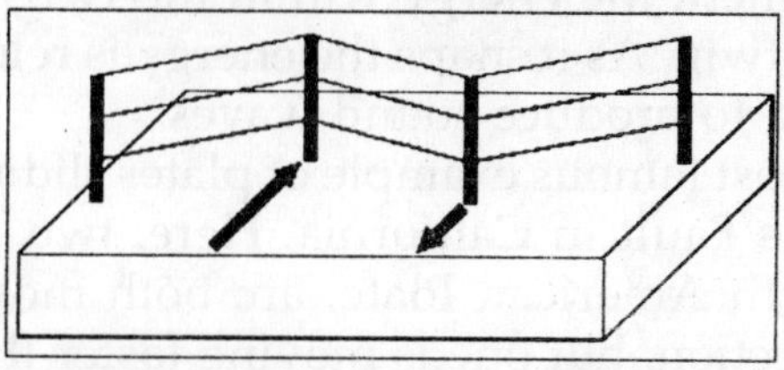

Fig. S-Waves Move Particles at 90° to the Wave's Direction.

L-Waves

Surface Waves (L-Waves) are low frequency transverse vibrations with a long wavelength. They are created close to the epicentre and can only travel through the outer part of the crust. They are responsible for the majority of the building damage caused by earthquakes.

This is because L Waves have a motion similar to that of waves in the sea. The ground is made to move in a circular motion, causing it to rise and fall as visible waves move across the ground. Together with secondary effects such as landslides, fires and tsunami these waves account for the loss of approximately 10,000 lives and over $100 million per year.

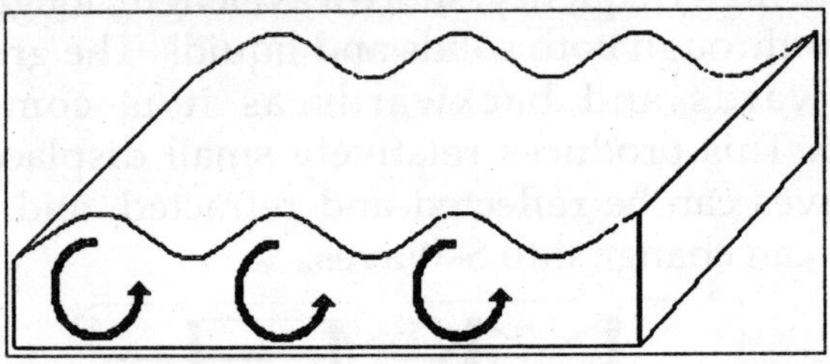

Fig. L-Waves Move Particles in a Circular Path.

Tectonic Earthquakes

Tectonic earthquakes are triggered when the crust becomes subjected to strain, and eventually moves. The theory of plate tectonics explains how the crust of the Earth is made of several plates, large areas of crust which float on the Mantle. Since these plates are free to slowly move, they can either drift towards each other, away from each other or slide past each other. Many of the earthquakes which we feel are located in the areas where plates collide or try to slide past each other.

The process which explains these earthquakes, known as *Elastic Rebound Theory* can be demonstrated with a green twig or branch. Holding both ends, the twig can be slowly bent. As it is bent, energy is built up within it. A point will be reached where the twig suddenly snaps. At this moment the energy within the twig has exceeded the *Elastic Limit* of the twig. As it snaps the energy is released, causing the twig to vibrate and to produce sound waves.

Perhaps the most famous example of plates sliding past each other is the San Andreas Fault in California. Here, two plates, the Pacific Plate and the North American Plate, are both moving in a roughly northwesterly direction, but one is moving faster than the other. The San Francisco area is subjected to hundreds of small earthquakes every year as the two plates grind against each other. Occasionally, as in 1989, a much larger movement occurs, triggering a far more violent 'quake'.

Major earthquakes are sometimes preceded by a period of changed activity. This might take the form of more frequent minor shocks as the rocks begin to move,called *foreshocks,* or a period of less frequent shocks as the two rock masses temporarily 'stick' and become locked together. Detailed surveys in San Francisco have shown that railway lines, fences and other longitudinal features very slowly become deformed as the pressure builds up in the rocks, then become noticeably offset when a movement occurs along the fault. Following the main shock, there may be further movements, called *aftershocks*, which occur as the rock masses 'settle down' in their new positions. Such aftershocks cause problems for rescue services, bringing down buildings already weakened by the main earthquake.

Volcanic Earthquakes

Volcanic earthquakes are far less common than Tectonic ones. They are triggered by the explosive eruption of a volcano. Given that not all volcanoes are prone to violent eruption, and that most are 'quiet' for the majority of the time, it is not surprising to find that they are comparatively rare.

When a volcano explodes, it is likely that the associated earthquake effects will be confined to an area 10 to 20 miles around its base, where as a tectonic earthquake may be felt around the globe.

The volcanoes which are most likely to explode violently are those which produce *acidic* lava. Acidic lava cools and sets very quickly upon contact with the air. This tends to chock the volcanic vent and block the further escape of pressure. For example, in the case of Mt Pelee, the lava solidified before it could flow down the sides of the volcano. Instead it formed a spine of solid rock within the volcano vent. The only way in which such a blockage can be removed is by the build up of pressure to the point at which the blockage is literally exploded out of the way. In reality, the weakest part of the volcano will be the part which gives way, sometimes leading to a sideways explosion as in the Mt St.Helens eruption.

When extraordinary levels of pressure develop, the resultant explosion can be devastating, producing an earthquake of considerable magnitude. When Krakatoa (Indonesia, between Java and Sumatra) exploded in 1883, the explosion was heard over 5000 km away in Australia. The shockwaves produced a series of tsunami (large sea waves), one of which was over 36m high; that's the same as four, two story houses stacked on top of each other. These swept over the coastal areas of Java and Sumatra killing over 36,000 people. By contrast, volcanoes producing free flowing *basic* lava rarely cause earthquakes. The lava flows freely out of the vent and down the sides of the volcano, releasing pressure evenly and constantly. Since pressure doesn't build up, violent explosions do not occur.

FAULT

In geology, a fault or fault line is a planar fracture in rock in which the rock on one side of the fracture has moved with respect to the rock on the other side. Large faults within the Earth's crust are the result of differential or shear motion and active fault zones are the causal locations of most earthquakes. Earthquakes are caused by energy release during rapid slippage along a fault. A fault that runs along the boundary between two tectonic plates is called a transform fault.

Since faults do not usually consist of a single, clean fracture, the term *fault zone* is used when referring to the zone of complex deformation that is associated with the fault plane. The two sides of a non-vertical fault are called the *hanging wall* and *footwall*. By definition, the hanging wall occurs above the fault and the footwall occurs below the fault. This terminology comes from mining. When working a tabular ore body the miner stood with the footwall under his feet and with the hanging wall hanging above him.

The creation and behaviour of faults, in both an individual small fault and within the greater fault zones which define the tectonic plates, is controlled by the relative motion of rocks on either side of the fault surface.

Because of friction and the rigidity of the rock, the rocks cannot simply glide or flow past each other. Rather, stress builds up in rocks and when it reaches a level that exceeds the strain threshold, the accumulated potential energy is released as strain, which is focused into a plane along which relative motion is accommodated — the fault.

Strain is both accumulative and instantaneous depending on the rheology of the rock; the ductile lower crust and mantle accumulates deformation gradually via shearing whereas the brittle upper crust reacts by fracture, or instantaneous stress release to cause motion along the fault. A fault in ductile rocks can also release instantaneously when the strain rate is too great. The energy released by instantaneous strain release is the cause of earthquakes, a common phenomenon along transform boundaries.

Mechanics

The creation and behaviour of faults, in both an individual small fault and within the greater fault zones which define the tectonic plates, is controlled by the relative motion of rocks on either side of the fault surface.

Because of friction and the rigidity of the rock, the rocks cannot simply glide or flow past each other. Rather, stress builds up in rocks and when it reaches a level that exceeds the strain threshold, the accumulated potential energy is released as strain, which is focused into a plane along which relative motion is accommodated — the fault.

Strain is both accumulative and instantaneous depending on the rheology of the rock; the ductile lower crust and mantle accumulates deformation gradually via shearing whereas the brittle upper crust reacts by fracture, or instantaneous stress release to cause motion along the fault. A fault in ductile rocks can also release instantaneously when the strain rate is too great. The energy released by instantaneous strain release is the cause of earthquakes, a common phenomenon along transform boundaries.

Microfracturing and AMR Theory

Microfracturing, or microseismicity, is sometimes thought of as a symptom caused by rocks under strain, where small-scale failures, perhaps on areas the size of a dinner plate or a small area, release stress under high strain conditions. It is only when sufficient microfractures link up into a large slip surface that a large seismic event or earthquake can occur.

According to this theory, after a large earthquake, the majority of the stress is released and the frequency of microfracturing is exponentially lower. A related theory, accelerating moment release (AMR), hypothesizes that the seismicity rate accelerates in a well-behaved way prior to large earthquakes, and may be a promising tool for earthquake prediction on the scale of days to years.

This is being increasingly used to predict rock failures within mines and applications are being attempted for the portions of faults within brittle rheological conditions. Similar behaviour is observed in the tremors preceding volcanic eruptions.

Slip, Heave, Throw

The sense of slip is defined by the relative movements of geological features present on either side of the fault plane and is a vector. The sense of slip defines the type of fault. This is distinct from the throw of the fault, which is the vertical offset. Heave is the measured horizontal offset of the fault.

The vector of slip can be qualitatively measured by fault bend folding, i.e. drag folding of strata on either side of the fault; the direction and magnitude of heave and throw can be measured only by finding common intersection points on either side of the fault. In practise it is usually only possible to find the slip direction of faults, and an approximation of the heave and throw vector.

Fault Types

Faults can be categorized into three groups based on the sense of slip. A fault where the main sense of movement (or slip) on the fault plane is vertical is known as a dip-slip fault. Where the main sense of slip is horizontal the fault is known as a transcurrent or strike-slip fault. Oblique-slip faults have significant components of both strike and dip slip.

For all naming distinctions, it is the orientation of the net dip and sense of slip of the fault which must be considered, not the present-day orientation, which may have been altered by local or regional folding or tilting.

Dip-Slip Faults

Dip-slip faults can be again classified into the types "reverse" and "normal". A normal fault occurs when the crust is extended. Alternatively such a fault can be called an extensional fault. The hanging wall moves downward, relative to the footwall. A downthrown block between two normal faults dipping towards each other is called a graben. An upthrown block between two normal faults dipping away from each other is called a horst. Low-angle normal faults with regional tectonic significance may be designated detachment faults.

A reverse fault is the opposite of a normal fault — the hanging wall moves up relative to the footwall. Reverse faults are indicative of shortening of the crust. The dip of a reverse fault is relatively steep, greater than 45°.

A thrust fault has the same sense of motion as a reverse fault, but with the dip of the fault plane at less than 45°. Thrust faults typically form ramps, flats and fault-bend (hanging wall and foot wall) folds. Thrust faults are responsible for forming nappes and klippen in the large thrust belts.

The fault plane is the plane that represents the fracture surface of a fault. Flat segments of thrust fault planes are known as *flats,* and inclined sections of the thrust are known as *ramps*. Typically thrust faults move *within* formations by forming flats, and climb up section with ramps.

Fault-bend folds are formed by movement of the hangingwall over a non-planar fault surface and are found associated with both extensional and thrust faults.

Faults may be reactivated at a later time with the movement in the opposite direction to the original movement (fault inversion). A normal fault may therefore become a reverse fault and vice versa.

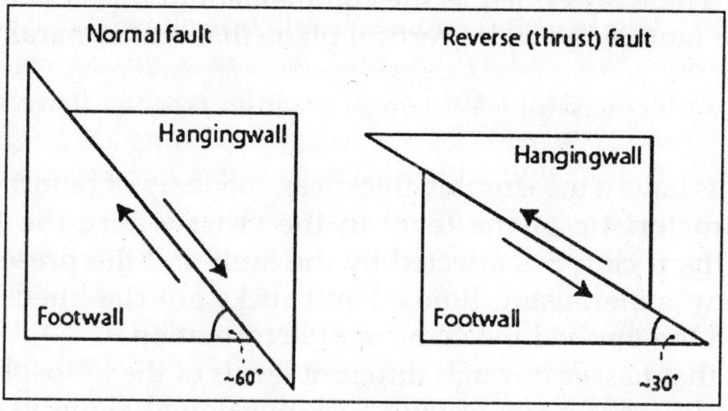

Fig. Cross-Sectional Illustration of Normal and Reverse dip-slip Faults.

Strike-Slip Faults

The fault surface is usually near vertical and the footwall moves either left or right or laterally with very little vertical motion. Strike-slip faults with left-lateral motion are also known as *sinistral* faults. Those with right-lateral motion are also known as *dextral* faults. A special class of strike-slip faults is the transform faults which are a plate tectonics feature related to spreading centers such as mid-ocean ridges.

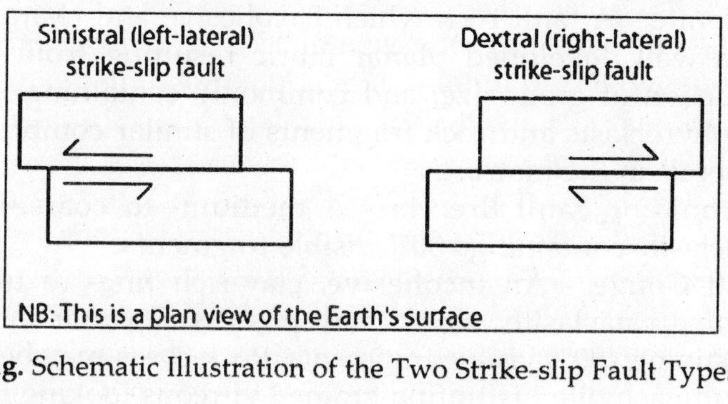

Fig. Schematic Illustration of the Two Strike-slip Fault Types.

Oblique-Slip Faults

A fault which has a component of dip-slip and a component of strike-slip is termed an oblique-slip fault. Nearly all faults will have some component of both dip-slip and strike-slip, so defining a fault as oblique requires both dip and strike components to be measurable and significant.

Some oblique faults occur within transtensional and transpressional regimes, others occur where the direction of extension or shortening changes during the deformation but the earlier formed faults remain active.

HADE: Hade is defined as the complement of dip, it is the angle between the fault plane and a vertical plane that strikes parallel to the fault.

Fault Rock

All faults have a measurable thickness, made up of deformed rock that is characteristic of the level in the crust where the faulting happened, the rock types affected by the fault and the presence and nature of any mineralising fluids. Fault rocks are classified by their textures and the implied mechanism of deformation.

A fault that passes through different levels of the lithosphere will have many different types of fault rock developed along its surface. Continued dip-slip displacement will tend to juxtapose fault rocks characteristic of different crustal levels, with varying degrees of overprinting. This effect is particularly clear in the case of detachment faults and major thrust faults.

The main types of fault rock are:

- Cataclasite - A fault rock which is cohesive with a poorly developed or absent planar fabric, or which is incohesive, characterised by generally angular clasts and rock fragments in a finer-grained matrix of similar composition.
- Mylonite - A fault rock which is cohesive and characterized by a well developed planar fabric resulting from tectonic reduction of grain size, and commonly containing rounded porphyroclasts and rock fragments of similar composition to minerals in the matrix
- Tectonic or Fault Breccia - A medium- to coarse-grained cataclasite containing >30% visible fragments.
- Fault Gouge - An incohesive, clay-rich fine- to ultrafine-grained cataclasite, which may possess a planar fabric and containing <30% visible fragments. Rock clasts may be present
- Pseudotachylite - Ultrafine-grained vitreous-looking material, usually black and flinty in appearance, occurring as thin planar veins, injection veins or as a matrix to pseudoconglomerates or breccias, which infills dilation fractures in the host rock.

- Clay smear clay-rich fault gouge formed in sedimentary sequences containing clay-rich layers which are strongly deformed and sheared into the fault gouge.

AFTERSHOCK

An aftershock is an earthquake that occurs after a previous earthquake (the main shock). An aftershock is in the same region of the main shock but is always of smaller magnitude strength. If an aftershock is larger than the main shock, the aftershock is redesignated as the main shock and the original main shock is redesignated as a foreshock. Aftershocks are smaller earthquakes formed as the crust around the displaced fault plane adjusts to the effects of the main shock.

Distribution of Aftershocks

Most aftershocks are located over the full area of fault rupture and either occur along the fault plane itself or along other faults within the volume affected by the strain associated with the main shock. Typically, aftershocks are found up to a distance equal to the rupture length away from the fault plane. The pattern of aftershocks helps confirm the size of area that slipped during the main shock. In the case of the 2004 Indian Ocean earthquake and the 2008 Sichuan earthquake the aftershock distribution shows that in both cases that the epicenter (where the rupture initiated) lies to one end of the final area of slip, implying strongly asymmetric rupture propagation.

Aftershock Size and Frequency with Time

Aftershocks tend to obey a number of empirical laws concerning magnitude and frequency.

Omori's Law

Aftershocks occur with a pattern that follows Omori's law. Omori's law, or more correctly the modified Omori's law, is an empirical relation for the temporal decay of aftershock rates. In 1894, Omori published his work on the aftershocks of earthquakes, in which he stated that aftershock frequency decreases by roughly the reciprocal of time after the main shock.

$$n(t) = \frac{K}{c+t}$$

where:

- $n(t)$ is the rate of earthquakes measured in a certain time t after the main shock,

- K is the amplitude, and
- c is the "time offset" parameter.

The modified version of Omori's law, now commonly used, was proposed by Utsu in 1961.

$$n(t) = \frac{K}{(c+t)^P}$$

where

- p modifies the decay rate and typically falls in the range 0.7–1.5.

According to these equations, the rate of aftershocks decreases quickly with time. The rate of aftershocks is proportional to the inverse of time since the mainshock. Thus whatever the odds of an aftershock are on the first day, the second day will have 1/2 the odds of the first day and the tenth day will have approximately 1/10th the odds of the first day (when p is equal to 1).

These patterns describe only the mass behaviour of aftershocks; the actual times, numbers and locations of the aftershocks are 'random', while tending to follow these patterns. As this is an empirical law values of the parameters are obtained by fitting to data after the mainshock occurred and they have no physical basis/meaning.

Bath's Law

The other main law describing aftershocks is known as Bath's Law and this states that the difference in magnitude between a main shock and its largest aftershock is approximately constant, independent of the main shock magnitude, typically 1.1-1.2 on the Moment magnitude scale.

Gutenberg-Richter Law

Aftershock sequences also typically follow the Gutenberg-Richter law of scaling, which refers to the relationship between the magnitude and total number of earthquakes in a region in a given time period. In summary, there are more small aftershocks and fewer large aftershocks.

Impact of Aftershocks

Aftershocks are dangerous because they are usually unpredictable, can be of a large magnitude, and can collapse buildings that are damaged from the mainshock. Bigger earthquakes have more and larger aftershocks and the sequences can last for years or even longer especially when a large event occurs in a seismically quiet area; the New Madrid Seismic Zone, where events still follow Omori's law from the mainshocks of 1811–1812. An aftershock sequence is deemed to have ended when the rate of seismicity drops back to a background level; i.e., no further decay in the number of events with time can be detected.

Foreshocks

Foreshocks have been defined as a tremor of at least 2.5 on the Richter-scale that take place prior to a mainshock tremor of 5.4 or greater on the Richter-scale. Many scientists hope to use foreshocks to predict upcoming earthquakes. In particular, the East Pacific Rise transform faults show foreshock activity before the main seismic event. Reviews of data of past events and their foreshocks showed that they have a low number of aftershocks and high foreshock rates compared to continental strike-slip faults.

EARTHQUAKE SWARM

Earthquake swarms are events where a local area experiences sequences of many earthquakes striking in a relatively short period of time. The length of time used to define the swarm itself varies, but the United States Geological Survey points out that an event may be on the order of days, weeks, or months.

They are differentiated from earthquakes succeeded by a series of aftershocks by the observation that no single earthquake in the sequence is obviously the main shock. Earthquake swarms are one of the events typically preceding eruptions of volcanoes.

One example is along the Cerro Prieto fault near Mexicali, BC in Mexico where over 500 quakes and aftershocks hit in February, 2008. Another is a swarm that's been dubbed "The Mogul earthquake sequence" that began in February 2008 near Reno, Nevada and continued for several months, ending in November 2008. Between February and April the swarm produced more than 1,000 quakes of small magnitude, although the largest measured 4.7.

Earthquake Storm

An earthquake storm is a recently proposed theory about earthquakes, where one triggers a series of other large earthquakes—within the same tectonic plate—as the stress transfers along the fault. This is similar to the idea of aftershocks, with the exception that they take place years apart.

These series of earthquakes can devastate entire countries or geographical regions. Possible events may have occurred during the end of the Bronze Age, and the latter part of the Roman Empire. It has been suggested that this is what may be occurring in modern day Turkey.

PLATE TECTONICS

The earth beneath our feet is not dead; it is constantly moving, driven by forces deep in its core. Nor is the planet's crust all of one piece; it is composed of numerous plates, which are moving steadily in relation to one another. This movement is responsible for all manner of phenomena, including earthquakes, volcanoes, and the formation of mountains. All these ideas, and many more, are encompassed in the concept of plate tectonics, which is the name for a branch of geologic and geophysical study and for a powerful theory that unites a vast array of ideas.

Plate tectonics works hand in hand with several other striking concepts and discoveries, including continental drift and the many changes in Earth's magnetic field that have taken place over its history. No wonder, then, that this idea, developed in the 1960s but based on years of research that preceded that era, is described as "the unifying theory of geology."

TECTONICS AND TECTONISM

The lithosphere is the upper layer of Earth's interior, including the crust and the brittle portion at the top of the mantle. Tectonism is the deformation of the lithosphere, and the term tectonics refers to the study of this deformation, including its causes and effects, most notably mountain building.

This deformation is the result of the release and redistribution of energy from Earth's core.

The interior of Earth itself is divided into three major sections: the crust, mantle, and core. The first is the uppermost division of the solid earth, representing less than 1% of its volume and varying in depth from 3 mi. to 37 mi. (5-60 km). Below the crust is the mantle, a thick, dense layer of rock approximately 1,429 mi. (2,300 km) thick.

The core itself is even more dense, as illustrated by the fact that it constitutes about 16% of the planet's volume and 32% of its mass. Composed primarily of iron and another, lighter element (possibly sulfur), it is divided between a solid inner core with a radius of about 760 mi. (1,220 km) and a liquid outer core about 1,750 mi. (2,820 km) thick.

Tectonism results from the release and redistribution of energy from Earth's interior. There are two components of this energy: gravity, a function of the enormous mass at the core, and heat from radioactive decay. Differences in mass and heat within the planet's interior, known as pressure gradients, result in the deformation of rocks.

DEFORMATION OF ROCKS

Any attempt to deform an object is referred to as stress, and stress takes many forms, including tension, compression, and shear. Tension acts to stretch a material, whereas compression—a type of stress produced by the action of equal and opposite forces, whose effect is to reduce the length of a material—has the opposite result. (Compression is a form of pressure.)

As for shear, this is a kind of stress resulting from equal and opposite forces that do not act along the same line. If a thick, hardbound book is lying flat and one pushes the front cover from the side so that the covers and pages are no longer perfectly aligned, this is an example of shear.

Under the effects of these stresses, rocks may bend, warp, slide, or break. They may even flow, as though they were liquids, or melt and thus truly become liquid. As a result, Earth's interior may manifest faults, or fractures in rocks, as well as folds, or bends in the rock structure. The effects of this activity can be seen on the surface in the form of subsidence, which is a depression in the crust, or uplift, which is the raising of crustal materials. Earthquakes and volcanic eruptions also may result.

There are two basic types of tectonism: orogenesis and epeirogenesis. *Orogenesis* is taken from the Greek words *oros* ("mountain") and *genesis* ("origin") and involves the formation of mountain ranges by means of folding, faulting, and volcanic activity. The Greek word *epeiros* means "mainland," and epeirogenesis takes the form of either uplift or subsidence. Of principal concern in the theory of plate tectonics, as we shall see, is orogenesis, which involves more lateral, as opposed to vertical, movement.

CONTINENTAL DRIFT

If one studies a world map for a period of time, one may notice something interesting about the shape of Africa's west coast and that of South America's east coast: they seem to fit together like pieces of a jigsaw puzzle. Early in the twentieth century, two American geologists, Frank Bursley Taylor and Howard Baker, were among the first scientists to point out this fact. According to Taylor and Baker, Europe, the Americas, and Africa all had been joined at one time. This was an early version of continental drift, a theory concerning the movement of Earth's continents.

Continental drift is based on the idea that the configuration of continents was once different than it is today, that some of the individual landmasses of today once were joined in other continental forms, and that the landmasses later moved to their present locations. Though Taylor and Baker were early proponents, the theory is associated most closely with the German geophysicist and meteorologist Alfred Wegener, who made the case for continental drift in *The Origin of Continents and Oceans*.

PANGAEA, LAURASIA, AND GONDWANALAND

According to Wegener, the continents of today once formed a single supercontinent called Pangaea, from the Greek words *pan* ("all") and *gaea* ("Earth"). Eventually, Pangaea split into two halves, with the northern continent of Laurasia and the southern continent of Gondwanaland, sometimes called Gondwana, separated by the Tethys Sea. In time, Laurasia split to form North America, the Eurasian landmass with the exception of the Indian subcontinent, and Greenland. Gondwanaland also split, forming the major southern landmasses of the world: Africa, South America, Antarctica, Australia, and India.

The Austrian geologist Eduard Suess and the South African geologist Alexander du Toit, each of whom contributed significantly to continental drift theory, were responsible for the naming of Gondwanaland and Laurasia, respectively. Suess preceded Wegener by many years with his theory of Gondwanaland, named after the Gondwana region of southern India.

There he found examples of a fern that, in fossilized form, had been found in all the modern-day constituents of the proposed former continent. Du Toit, Wegener's contemporary, was influenced by continental drift theory and improved on it greatly.

FORMATION OF THE CONTINENTS

Today continental drift theory is accepted widely, in large part owing to the development of plate tectonics, "the unifying theory in geology." We examine the evidence for continental drift, the arguments against it, and the eventual triumph of plate tectonics in the course of this essay. Before going on, however, let us consider briefly the now-accepted timeline of events described by Wegener and others.

About 1,100 million years ago (earth scientists typically abbreviate this by using the notation 1,100 Ma), there was a supercontinent named Rodinia, which predated Pangaea. It split into Laurasia and Gondwanaland, which moved to the northern and southern extremes of the planet, respectively. Starting at about 514 Ma, Laurasia drifted southward until it crashed into Gondwanaland about 425 Ma. Pangaea, surrounded by a vast ocean called Panthalassa, formed approximately 356 Ma.

In the course of Pangaea's formation, what is now North America smashed into northwestern Africa, forming a vast mountain range. Traces of these mountains still can be found on a belt stretching from the southern United States to northern Europe, including the Appalachians.

As Pangaea drifted northward and smashed into the ocean floor of Panthalassa, it formed a series of mountain ranges from Alaska to southern South America, including the Rockies and Andes. By about 200 Ma, Pangaea began to break apart, forming a valley that became the Atlantic Ocean. But the separation of the continents was not a "neat" process: today a piece of Gondwanaland lies sunken beneath the eastern United States, far from the other landmasses to which it once was joined.

By about 152 Ma, in the late Jurassic period, the continents as we know them today began to take shape. By about 65 Ma, all the present continents and oceans had been formed for the most part, and India was drifting north, eventually smashing into southern Asia to shape the world's tallest mountains, the Himalayas, the Karakoram Range, and the Hindu Kush. This process is not finished, however, and geologists believe that some 250-300 million years from now, Pangaea will re-form.

EVIDENCE AND ARGUMENTS

As proof of his theory, Wegener cited a wide variety of examples, including the apparent fit between the coastlines of South America and western Africa as well as that of North America and northwestern Africa. He also noted the existence of rocks apparently gouged by glaciers in southern Africa, South America, and India, far from modern-day glacial activity.

Fossils in South America matched those in Africa and Australia, as Suess had observed. There were also signs that mountain ranges continued between continents—not only those apparently linking North America and Europe but also ranges that seemed to extend from Argentina to South Africa and Australia.

By measurements conducted over a period of years, Wegener even showed that Greenland was drifting slowly away from Europe, yet his theory met with scorn from the geoscience community of his day. If continents could plow through oceanic rock, some geologists maintained, then they would force up mountains so high that Earth would become imbalanced. As for his claim that matching fossils in widely separated regions confirmed his theory of continental drift, geologists claimed that this could be explained by the existence of land bridges, now sunken, that once had linked those areas. The apparent fit between present-day landmasses could be explained away as coincidence or perhaps as evidence that Earth simply was expanding, with the continents moving away from one another as the planet grew.

PLATE TECTONICS

Though Wegener was right, as it turned out, his theory had one major shortcoming: it provided no explanation of exactly *how* continental drift had occurred. Even if geologists had accepted his claim that the continents are moving, it raised more questions than answers. A continent is a very large thing simply to float away; even an aircraft carrier, which is many millions of times lighter, has to weigh less than the water it displaces, or it would sink like a stone. In any case, Wegener never claimed that continents floated. How, then, did they move?

The answer is plate tectonics, the name both of a theory and of a specialization of tectonics. As an area of study, plate tectonics deals with the large features of the lithosphere and the forces that fashion them. As a theory, it explains the processes that have shaped Earth in terms of plates (large movable segments of the lithosphere) and their movement. Plate tectonics theory brings together aspects of continental drift, seafloor spreading, seismic and volcanic activity, and the structures of Earth's crust to provide a unifying model of Earth's evolution.

THE PIECES COME TOGETHER

In 1962 the United States geologist Harry Hammond Hess introduced a new concept that would prove pivotal to the theory of plate tectonics: seafloor spreading, the idea that seafloors crack open along the crest of mid-ocean ridges and that new seafloor forms in those areas. According to Hess, a new floor forms when molten rock called magma rises up from the asthenosphere, a region of extremely high pressure underlying the lithosphere, where rocks are deformed by enormous stresses. The magma wells up through a crack in a ridge, runs down the sides, and solidifies to form a new floor.

Three years later, the Canadian geologist John Tuzo Wilson coined the term plates to describe the pieces that make up Earth's rigid surface. Separated either by the mid-ocean rifts identified earlier by Heezen or by mountain chains, the plates move with respect to one another. Wilson presented a model for their behaviour and established a global pattern of faults, a sort of map depicting the movable plates. The pieces of a new theory were forming, but as yet it had no name.

That name appeared in 1967, when D. P. Mackenzie of England and R. L. Parker of the United States introduced the term plate tectonics. They maintained that the surface of Earth is divided into six major as well as seven minor movable plates and compared the continents to enormous icebergs—much as Wegener had described them half a century earlier. Subsequent geologic research has indicated that there may be as many as nine major plates and as many as 12 minor ones.

To test these emerging ideas, the U.S. National Science Foundation authorized a research voyage by the vessel *Glomar Challenger* in 1968. On their first cruise, through the Gulf of Mexico and the Atlantic, the *Challenger*'s scientific team collected sediment, fossil, and crust samples that confirmed the basics of seafloor spreading theory.

These results led to new questions regarding the reactions between rocks and the heated water surrounding them, spawning new research and necessitating additional voyages. In the years that followed, the *Challenger* made more and more cruises, its scientific teams collecting a wealth of evidence for the emerging theory of plate tectonics.

REAL-LIFE APPLICATIONS

Early Evidence of Plate Tectonics

No single person has been as central to plate tectonics as Wegener was to continental drift or as the English naturalist Charles Darwin was to evolution. The roots of plate tectonics lie partly in the observations of Wegener and other proponents of continental drift as well as in several discoveries and observations that began to gather force in the third quarter of the twentieth century.

During World War II, submarine warfare necessitated the development of new navigational technology known as sonar (*SO* und *N* avigation *A* nd *R* anging). Sonar functions much like radar, but instead of using electromagnetic waves, it utilizes ultrasonic, or high-frequency, sound waves projected through water. Sonar made it feasible for geologists to study deep ocean basins after the war, making it possible for the first time in history to map and take samples from large areas beneath the seas. These findings raised many questions, particularly concerning the vast elevation differences beneath the seas.

Ewing and the Mountains Under the Ocean

One of the first earth scientists to notice the curious aspects of underwater geology was the American geologist William Maurice

Ewing, who began his work long before the war. He had gained his first experience in a very practical way during the 1920s, as a doctoral student putting himself through school. Working summers with oil exploration teams in the Gulf of Mexico off the coast of Texas had given him a basic understanding of the subject, and in the following decade he went to work exploring the structure of the Atlantic continental shelf and ocean basins.

His work there revealed extremely thick sediments covering what appeared to be high mountainous regions. These findings sharply contradicted earlier ideas about the ocean floor, which depicted it as a flat, featureless plain rather like the sandy-bottomed beaches found in resort areas. Instead, the topography at the bottom of the ocean turned out to be at least as diverse as that of the land above sea level.

Heezen and the Rift Valley

During the 1950s, a team led by another American geologist, Bruce Charles Heezen, worked on developing an overall picture of the ocean basin's topography. Earlier work had identified a mountain range running the length of the Atlantic, but Heezen's team discovered a deep valley down the middle of the chain, running parallel to it. They described it as a rift valley, a long trough bounded by two or more faults, and compared it to a similar valley in eastern Africa.

Around the same time, a group of transatlantic telephone companies asked Heezen to locate areas of possible seismic or earthquake activity in the Atlantic. Phone company officials reasoned that if they could find the areas most likely to experience seismic activity, they could avoid placing their cables in those areas. As it turned out, earthquakes tended to occur in exactly the same region that Heezen and his team had identified as the rift valley.

The Plates and Their Interactions

The most significant plates that make up Earth's surface are as follows:
Selected Major Plates

- North American (almost all of North America and Mexico, along with Greenland and the northwestern quadrant of the Atlantic).
- South American (all of South America and the southwestern quadrant of the Atlantic).
- African (Africa, the southeastern Atlantic, and part of the Indian Ocean).
- Eurasian (Europe and Asia, excluding the Indian subcontinent, along with surrounding ocean areas).

- Indo-Australian (India, much of the Indian Ocean, Australia, and parts of the Indonesian archipelago and New Zealand).
- Antarctic (Antarctica and the Antarctic Ocean).

In addition to these plates, there are several plates that while they are designated as "major" are much smaller: the Philippine, Arabian, Caribbean, Nazca (off the west coast of South America), Cocos (off the west coast of Mexico), and Juan de Fuca (extreme western North America). Japan, one of the most earthquake-prone nations in the world, lies at the nexus of the Philippine, Eurasian, and Pacific plates.

Movement of the Plates

One of the key principles of geology is uniformitarianism: the idea that processes occurring now also occurred in the past. The reverse usually is also true; thus, as we have noted, the plates are still moving, just as they have done for millions of years. Thanks to satellite remote sensing, geologists are able to measure this rate of movement.

Not surprisingly, its pace befits the timescale of geologic, as opposed to human, processes: the fastest-moving plates are careening forward at a breathtaking speed of 4 in. (10 cm) per year. The ground beneath Americans' feet is drifting at the rate of 1.2 in. (3 cm) every year, which means that in a hundred years it will have shifted 10 ft. (3 m).

When Plates Interact

Plates interact by moving toward each other (convergence), away from each other (divergence), or past each other (transform motion). Convergence usually is associated with subduction, meaning that one plate is forced down into the mantle and eventually undergoes partial melting. This typically occurs in the ocean, creating a depression known as an oceanic trench. Divergence results in the separation of plates and most often is associated either with seafloor spreading or the formation of rift valleys.

There are three types of plate margins, or boundaries between plates, depending on the two types of crusts that are interacting: oceanic with oceanic, continental with continental, or continental with oceanic. The rift valleys of the Atlantic are an example of an oceanic margin where divergence has occurred, while oceanic convergence is illustrated by a striking example in the Pacific. There, subduction of the Philippine Plate by the Pacific Plate has created the Mariana Trench, which at 36,198 ft. (10,911 m) is the deepest depression on Earth.

When continental plates converge, neither plate subducts; rather, they struggle against each other like two warriors in a fight to the death, buckling, folding, and faulting to create huge mountain ranges. The convergence of the Indo-Australian and Eurasian plates has created the *highest* spots on Earth, in the Himalayas, where Mount Everest (on the Nepal-Indian border) rises to 29,028 ft. (8,848 m). Continental plates also may experience divergence, resulting in the formation of seas. An example is the Red Sea, formed by the divergence of the African and Arabian plates.

Given these facts about the interactions of oceanic and continental plates with each other, what occurs when continental plates meet oceanic ones is no surprise.

In this situation, the oceanic plate meeting the continental plate is like a high-school football player squaring off against a National Football League pro tackle. It is no match: the oceanic plate easily subducts. This leads to the formation of a chain of volcanoes along the continental crust, examples being the Cascade Range in the U.S. Pacific Northwest (Juan de Fuca and Pacific plates) or the Andes (South American and Nazca plates).

Transform margins may occur with any combination of oceanic or continental plates and result in the formation of faults and earthquake zones. Where the North American Plate slides against the Pacific Plate along the California coast, it has formed the San Andreas Fault, the source of numerous earthquakes, such as the dramatic San Francisco quakes of 1906 and 1989 and the Los Angeles quake of 1994.

Paleomagnetism

As noted earlier, plate tectonics brings together numerous areas of study in the geologic sciences that developed independently but which came to be seen as having similar roots and explanations. Among these disciplines is paleomagnetism, an area of historical geology devoted to studying the direction and intensity of magnetic fields in the past, as discerned from the residual magnetization of rocks.

Earth has a complex magnetic field whose principal source appears to be the molten iron of the outer core. In fact, the entire planet is like a giant bar magnet, with a north pole and a south pole. It is for this reason that the magnetized material in a compass points north; however, Earth's magnetic north pole is not the same as its geographic north pole. It so happens that magnetic north lies in more or less the same direction as geographic north, but as geologists in the mid-nineteenth century discovered, this has not always been the case.

In 1849 the French physicist Achilles Delesse observed that

magnetic minerals tend to line up with the planet's magnetic field, pointing north as though they were compass needles. Nearly 60 years later, however, another French physicist, Bernard Brunhes, noted that in some rocks magnetic materials point *south*. This suggested one of two possibilities: either the planet's magnetic field had reversed itself over time, or the ground containing the magnetized rocks had moved. Both explanations must have seemed far-fetched at the time, but as it turned out, both are correct.

Earth's magnetic field has shifted, meaning that the magnetic north and south poles have changed places many times over the eons. In addition, the magnetic poles have wandered around the southern and northern portions of the globe: for instance, whereas magnetic north today lies in the frozen islands to the north of Canada, at about 300 Ma it was located in eastern Siberia.

The movement of magnetic rocks on Earth's surface, however, has turned out to be too great to be explained either by magnetic shifts or by regional wandering of the poles. This is where plate tectonics and paleomagnetism come together.

Confirmation of Plate Tectonic Theory

Rocks in Alaska have magnetic materials aligned in such a way that they once must have been at or near the equator. In addition, the orientation of magnetic materials on South America's east coast shows an affinity with that of similar materials on the west coast of Africa. In both cases, continental drift, with its driving mechanism of plate tectonics, seems the only reasonable explanation.

Thus, paleomagnetic studies have served to confirm the ideas of continental drift and plate tectonics, while research conducted at sea bolsters seafloor spreading theory. Using devices called magnetometers, geologists have found that the orientation of magnetic minerals on one side of a rift mirrors that of materials on the other side. This suggests that the new rock on either side of the rift was formed simultaneously, as seafloor spreading theory indicates.

Earthquakes and Volcanoes

Several findings relating to earthquakes and volcanic activity also can be explained by plate tectonics. If one follows news stories of earthquakes, one may begin to wonder why such places as California or Japan have so many quakes, whereas the northeastern United States or western Europe have so few. The fact is that earthquakes occur along belts, and the vast majority of these belts coincide with the boundaries between Earth's major tectonic plates.

The same is true of volcanoes, and it is no mistake that places famous for earthquakes—the Philippines, say, or Italy—often also are known for their volcanoes. Although they are located near the centre of the Pacific Plate, the islands of Hawaii are subject to plate movement, which has helped generate the volcanoes that gave those islands their origin.

At the southern end of the island chain, many volcanoes are still active, while those at the northern end tend to be dormant. The reason is that the Pacific Plate as a whole is moving northward over a stationary lava source in the mantle below Hawaii. The southern islands remain poised above that source, while the northern islands have moved away from it.

Chapter 12

Volcanic Erruption

The structures and relationships of igneous masses can be studied only in regions where these masses have been uncovered by erosion of the rocks which formerly covered them, and their origin can be determined only from the available facts. We shall first consider those masses the formation of which are open to observation or which have been formed so recently that there can be no doubt as to their origin.

CLASSIFICATION OF VOLCANOES

Volcanoes are found in many parts of the world at the present time, but their most extensive distribution is around the borders of the Pacific, where they form a "chain of fire" which surrounds that ocean. The map on the opposite page shows this and their occurrence elsewhere. Along the western border of the Pacific more than 150 active volcanoes are distributed through a length of 16,000 kilometers, while about 100 border the eastern margin of that ocean from the Aleutian Islands to Tierra del Fuego.

Typical and perfect examples of the former are Fujiyama in Japan and Mayon in the Philippines, while Bogosloff and Popocatepetl may serve as examples of the latter. Within the Pacific and arising from its floor are other active volcanoes, such as those of the Hawaiian Islands, near its very centre, and those of the southwest coral reef regions.

On the Atlantic borders, active volcanoes are not common, being chiefly confined to the Antillean region in the west, the Icelandic in the north, and the Azores, Canaries, and Cape Verde Islands in the east. These are in reality local or isolated groups of volcanoes, except those of Iceland, which are the last manifestations of volcanic activity in a vast volcanic field which extends from Greenland on the west to Siberia on the east, and which began far back in the geological history of the region. In the Mediterranean are three important fields of volcanic activity known since ancient times: that around Naples, that of Sicily and its islands, and that of the Grecian archipelago. The Naples volcanic field comprises Vesuvius with Monte Somma on the east of

Naples, the Phlegræan volcanic field with a number of vents including Monte Nuovo, the Solfatara, the Pozzuoli district, etc., on the west, and the volcanic island of Ischia on the southwest, besides many smaller islands, including Capri on the south.

The Sicilian region includes Mount Etna in the northeastern part of that island, and the volcanoes of the Lipari or Æolian group of islands north of Sicily, among which are Vulcano and Stromboli, the latter called the "Lighthouse of the Mediterranean," because of the flashes of its explosions, given off at intervals of from 1 to 20 minutes.In the Grecian archipelago the island of Santorin has been a scene of volcanic activity for more than 2000 years.

Volcanoes are also active around the Indian Ocean, especially on the islands which form its eastern margin. Active volcanoes exist in the great "Rift Valley" of East Africa, in which lie many of the large lakes of that region, and recently extinct volcanoes occur along the northern margin of the Indian Ocean in Arabia, while more ancient volcanic activities are recorded in India.

Active, Dormant, and Extinct Volcanoes

Volcanoes may be classed as active when eruptions are occurring at intervals; as *dormant*, when they have been quiescent for centuries, though the possibility of a new eruption exists; and as extinct, when volcanic activity has entirely ceased and the volcano is undergoing destruction by atmospheric agencies. There is of course every gradation between these; some that would be classed as dormant may never be revived, while others, after a long sleep, burst forth again, as in the case of Vesuvius at the beginning of the Christian Era.

It should be clearly understood that volcanism is not a phenomenon restricted to the present, but that volcanic activities, often on a much grander scale than those of to-day, have been going on during all the periods of the earth's history. Many of the older volcanoes and their products, after suffering more or less destruction by erosion, were buried under more recent deposits, and have been reëxposed only in part as the result of the latest phases of erosion.

Monte Nuovo

This volcano arose in the Phlegræan fields west of Naples on Sunday, September 29, 1538, beginning about one o'clock in the morning. It appeared mainly on the site of the ancient Lake Lucrinus, which itself was regarded as the crater of an older but extinct preëxisting volcano, filled with water. Incandescent gases burst open the earth, and within a week a cone made up of ejected ashes, cinders, and large stones, but no lava, was built up to a height of 440 feet above the level of the sea. The depth of the crater is 421 feet, reaching within 19 feet of the sealevel, and its basal circumference about 8000 feet. The eruption was preceded on the day and night before by about twenty earthquake shocks in Pozzuoli and the neighborhood, and after the eruption it was found that the sea had receded for some distance, owing to the elevation of the region, and the strand was covered with quantities of dead fish, as well as dead birds.

The cone of Monte Nuovo is very regular, and is composed of layers of pumaceous fragments and ashes, and of trachytic blocks, the whole partly consolidated and dipping away from the crater at angles of 26 to 30 degrees. Fragments of marine shells and Roman bricks were also included in the beds as a result of the disturbance of the region by the explosions.

Jorullo

This cone arose on the night of September 28, 1759, at a point 35 miles distant from any then existing volcano, and 120 miles from the sea. It appeared on a level plain 2000 to 3000 feet above the sea in the state of Michoacan, Mexico. A northeasterly fissure opened in the midst of the sugar and indigo fields of this plain, and ashes and rocks were thrown to a great height, and on falling these built up six conical hills on the line of the chasm, the smallest 300 feet high, while Jorullo itself was built up to 1600 feet above the plain, or 4265 feet above sea-level. Great streams of basaltic lava were poured forth from Jorullo, these including fragments of granitic rock.

The ejection did not cease until February, 1760. Twenty years later the lava was still hot enough, a few inches below the surface, to light a cigar. This eruption, too, was preceded by earthquakes and subterranean rumblings, which began in the June preceding. There has been no eruption in this region since.

Camiguin

This volcano also started from a fissure in a level plain on one of the small islands north of Luzon in the Philippines. Beginning in 1871, it continued to be active for four years, by which time it had reached a height of about 1800 feet.

Sabrina Island

This submarine volcanic cone appeared above the waters of the Atlantic in the Azores group, off the coast of St. Michaels, on June 13, 1811, and rose to a height of about 300 feet above the sea, gaining a circumference of about a mile. Being, however, largely made of unconsolidated material, it has since been washed away again. As observed from the nearest cliff on St. Michaels, the explosions resembled a mixed discharge of cannon and musketry and were accompanied by a great abundance of lightning.

Graham Island

This island, which existed for only about three months, rose as a submarine cinder cone in 1831 in the Mediterranean between the southwest coast of Sicily and that projecting part of the African coast where ancient Carthage stood. A few years before the appearance of the island, soundings at this locality showed a depth of water of 100 fathoms. Premonitory shocks were felt on June 28 over the spot and on the adjoining coast of Sicily. About July 10, a column of water 60 feet high and 800 yards in circumference was seen rising from the sea, followed soon after by dense clouds of steam which rose 1800 feet. Eight days later the same observer noted at this spot a small island 12 feet high and with a crater at its centre, from which volcanic matter and immense clouds of vapour were ejected, while the sea round about was covered with floating cinders and dead fish.

By the end of July the island had become from 50 to 90 feet in height and three fourths of a mile in circumference, while on August 4 it was reported above 200 feet high and three miles in circumference. After this it began to diminish in size, owing to the erosion by the sea, decreasing to two miles in circumference by August 25, and to three fifths of a mile and a maximum height of 107 feet by September 3, when the crater was about 780 feet in circumference.

On September 29 the island was reduced to a circumference of only about 700 yards, and toward the close of October it had disappeared except for a small point of sand and scoriæ. By the commencement of 1832 there was only a shoal, with a mass of igneous rock which apparently filled the centre of the vent. The appearance and supposed structure of this island and of the cone, which rose thus about 800 feet above the sea floor reproduced from Lyell.

The lava core or neck, which probably never appeared above sea-level, now form the highest part of the submerged remnant of the volcano. The fragmental material included, besides volcanic ash and cinders, blocks of limestone, dolomite, and sandstone.

Several other submarine eruptions of this type have been recorded from various sections of the Atlantic and from the Mediterranean.

CHARACTERISTIC OF VOLCANOES

Of the many modern volcanoes, a number which have been

observed for long periods of time may be described somewhat in detail so that the student will get a grasp of the essentials of the structure and activities of volcanoes. We shall begin with a type in which liquid lava is the chief product of eruption, then consider types in which both lava and fragmental material are produced, and enter into the building of the cone, and finally consider some examples which are purely explosive, with the production of only fragmental material.

Before proceeding, the student should clearly understand that the hill or mountain which constitutes the volcano is built up around an opening or vent from the material ejected from this vent, and that it does not represent an upris ing or an upheaval of a part of the earth's surface, as was at one time supposed to be the case. Three kinds of material are ejected from volcanic vents; of these, two or all may be present. The first type comprises gases and *water vapors* mingled with fumes of other substances; this is always present.

The second is liquid magma or lava, and the third consists of the shattered lava and shattered rocks resulting from the explosive activities, comprising masses of all sizes, from large lava balls or bombs, more or less spherical or elliptical from their rotary motion through the air, to fine lava particles or *lapilli,* and fragments or blocks of older igneous or other rocks, and dust produced by the shattering of these. In general we speak of such material as volcanic ashes and cinders. The eruptive activities range from the quiet upwelling or bubbling-up of liquid lava to the most violent explosion due to the sudden expansion of the gases and vapors.

Kilauea of the Hawaiian Islands

The entire group of the Hawaiian Islands in the midPacific is a series of volcanic cones built up from the sea-bottom in former times. Most of these volcanoes are now extinct, and many of them are undergoing erosion; but a large, active crater, that of Kilauea (4000 feet above sea-level), exists on the east side of the volcanic mountain of Mauna Loa (13,675 ft. high) and about twenty miles from its summit, which is also marked by an active crater.

The crater pit of Kilauea is rudely oval in form, with a circumference of about nine miles, and its floor is formed by a rough stony crust of solidified lava — resting upon a vast column of molten rock which arises from an unknown depth in the crust of the earth. In places, this floor is broken by lakes of liquid lava, red to white hot, and set into boiling activity by the ebullition of gases. From cracks in the floor and on its margin jets of lava are frequently projected to great heights, and some of this material, blown like spray by the wind, is drawn out into slender, hair-like fibers. This has become known as Pele's hair, so named after the goddess of the mountains.

The margin of the crater is formed by a precipitous cliff which varies in height from time to time. This is due to the fact that the lava column which supports the floor rises as the pressure increases below, until a point is reached where the wall may be ruptured, and the liquid lava flows out, whereupon the column dimin shes, carrying the floor downward, until it may be 700 feet below the edge of the crater rim. The rate of rising may be as much as 100 feet in a year, but in modern times the lava has not overflowed the rim, but issues from lateral fissures due to cracking or fusion, the floor approaching only to within 300 feet of the edge of the crater.

During the eruption of 1840 the lava first appeared on the side of the mountain five miles from the main crater, after which it issued at successively lower levels.

Similar conditions exist in the crater of Mauna Loa proper, but here the top of the lava column is nearly 10,000 feet above that of Kilauea.

The lava of these volcanoes is very liquid, being of an extremely basic character and forming basaltic rocks on cooling. Because of its liquidity, it will continue to flow for a long time; some of the lava flows of Hawaii are thirty miles in length. When it reaches a cliff, the lava cascades over it like a waterfall. The surface of such lava streams often shows local wave-like advances, which produce the appearance of a series of crushed pillows piled one against the other. This type of surface form is known in Hawaii as Pahoehoe, and it is seen in older lavas of this type in many regions of the world. A ropy surface, having the appearance of coils of heavy rope, is also commonly produced.

As the lava stream moves along, it sweeps away forests in its course, and carries away masses of rock and soil covered with vegetation.

Sometimes a stream will part around a mass of such rock and, reuniting, enclose it as an island. On reaching the sea, the lava plunges into it with loud detonations and becomes shivered into millions of particles of glass, which may be thrown in clouds into the air. The light from such an eruption has been visible for over a hundred miles at sea, and at a distance of forty miles fine print could be read at midnight.

When the crust of the lava stream has cooled, the interior mass, still in a molten condition, will flow on, leaving a tunnel behind. Such tunnels are common in Hawaii, and their roofs are frequently incrusted with lava pendants or stalactites up to 20 or 30 inches in length, while on the floor corresponding drip mounds or stalagmites of lava are found. Small "spatter cones" may also arise on its surface. The form of these basic lava volcanoes is very characteristic, being a very flat cone of vast dimensions. As they arise from the seabottom, only a small part is visible, the total height of Mauna Loa being more than 30,000 feet, although considerably less than half of this height is seen. The summit is nearly flat for several square miles, and the slopes of the sides do not average more than seven degrees.

CLASSIFICATION OF VOLCANOES

We have now seen something of the mode of eruption of several distinct types of volcanoes in various parts of the earth. According to their mode of eruption, we may classify them as the quiet type on the one extreme, represented by the welling up and pouring out of liquid lavas, as in Kilauea, and the explosive type, represented by Krakatoa and Bandai-San, where shattering of rock material occurs, but without the outpouring of liquid lava.

In the milder examples of this type a cinder cone is built up. Between these two stand the types which show both kinds of eruption, with the result that beds of volcanic ashes and lapilli alternate with beds of solidified lava. Here we place Vesuvius and Etna, explosive eruptions being more marked in the former and lava eruptions in the latter.

Comparing the form of the cinder cone with that of a pure lava cone. The former, illustrated by the wonderfully perfect cinder cone of Mayon in the Philippines (Frontispiece) has steep slopes, the angle being determined by the nature and coarseness of the fragmental material. The lava cone, on the other hand, especially that composed of basic lava, is broad and relatively low, though the crater may be situated at a great height above the base. The slopes are very gentle, and the top generally a plateau. This is illustrated by Mauna Loa in the Hawaiian group.

GEOLOGICAL AGE OF VOLCANOES AND LAVA FLOWS

The geological age of a volcano can be determined from the age of the associated formations. It is obvious that a lava flow is always younger than the formation upon which it rests, and older than that which covers it. Lava sheet which rests upon river gravels of Pleistocene age and is therefore younger than these. The extensive erosion which it has suffered, indicates that it is probably of late Pleistocene age.

STRUCTURAL CHARACTERS OF VOLCANOES

THE structural character of volcanoes is revealed in cones that have become extinct, for in these the parts are not only more easily accessible, but dissection has often revealed the internal character as well.

EXTINCT VOLCANOES

There are many regions where volcanoes have been active in the recent geological past, and in such cases enough of the form and character of the volcanoes, now extinct, is still retained to enable one to recognize them readily.

Such recently extinct and partly dissected volcanoes are not only of interest as showing former distribution of volcanic activity, but they have an added value because their erosion has revealed many features which in an active volcano are not open to view. Thus the study of the recently extinct volcanoes supplements that of active ones.

Extinct Volcanoes of Central France

One of the most notable fields of former volcanic activity, and one that has played a prominent part in the history of the science of volcanology, lies in the central part of the great area of crystalline and younger rocks which makes up the so-called *Massif Central* of France, and which is bounded on the north by the Paris basin of Mesozoic and Tertiary rocks, on the southwest by the basin of the Garonne, and on the east by the valleys of the Rhone and Saône.

The centre of this massif is dissected by the river Allier, which flows north into the Loire, and it is along the western border of the Allier valley, which is bounded by a fault, that the main volcanic district is located, while a second one lies on the southeast, in the region of the headwaters of the Allier and the Loire.

The younger eruptive rocks belong to several geological epochs, namely the Pleistocene, the Pliocene, and the Miocene. Of these the youngest, of Pleistocene age, form the famous Chaîne des Puys, a group of extinct volcanoes which still retain to a remarkable degree their form and general character. South of this lies the volcanic district of Mont Dore, and still farther south that of Cantal — both of Pliocene age. To the east of the latter and between the Allier and the Loire is the Chaîne du Velay, of Pliocene basalts, and east of this a series of Miocene eruptive hills

The Chaîne des Puys

This is best approached from ClermontFerrand, which lies to the east of the highest of these old volcanic hills, the Puy de Dôme, from the summit of which an inspiring panorama of these silent volcanic cones, some sixty in number, may be seen. They extend for a distance of about 90 kilometers north from the Mont Dore region. The Puy de Dôme itself is not a perfect volcano, but is formed by a central dike or neck of trachytic rock (called by the French domite), but the great majority of the cones show each their cup-shaped crater at the top, and produce a volcanic topography rivaling that of the surface of the moon.

The lava of these volcanoes consists of olivine basalts and andesites, with abundant slags, scoriæ, and pumaceous material all rich in ferro-magnesian minerals, and containing from 50 to 58 per cent of silica. The outpourings belong to the middle and later Pleistocene time. Besides the perfect crater cones there are, however, great intumescences, or dome-like blisters of trachytic rock without craters, and these represent the type of up-swellings of viscous acidic lavas.

The Massif of Mont Dore

This mass, south of the Puys, represents the remnant of a great volcano, active during Pliocene time, but since then partly destroyed by erosion. While this makes possible the detailed study of the various parts of the volcano, and gives us an opportunity to observe the successive eruptions and their effects, it also makes a more difficult problem for the student to comprehend, for it must be borne in mind that the many peaks and prominences now seen are not separate volcanoes as in the Puy district, but erosion remnants of larger masses, and the imagination must be drawn upon to restore again what is missing and so get a of the whole volcano as it was during its prime.

The relationships of the different volcanic rocks which make up this volcano as would be obtained were the entire massif cut through its highest part (Sancy) like a round cake cut through the middle, and half of it removed so that the cut side of the other half is visible. Such a *cross-section,* as it is called, is built up from innumerable local observations which are then connected by logical inferences.

This section shows that the foundation of the volcano is ancient granite and gneiss, and that it was built up on an almost level surface (a peneplane) which had previously been cut across the old foundation rocks by natural agencies. Upon this floor lies a mass of andesitic tuff, the product of the first eruptive activity (upper Miocene). Then (early Pliocene) followed an eruption of trachyte porphyries through the Sancy vents and that of the Capucin, the flows of which extended laterally for some distance, finally thinning away. The next eruptions (middle Pliocene) were those of the Puy de Pailleret and the Puy de Cliergue on each side of Sancy, these lavas on cooling forming hornblendic andesites.

Then came the late Pliocene eruption of the Plateau basalts, which cut and rest upon the others, and this was followed by local Pleistocene eruptions of the basalts and the formation of cinder cones which correspond to those of the Puy Chain. The succession of events is clearly indicated by the relationships, especially the superposition of the various lavas and pyroclastic products, and it will be observed that the eruptions proceeded from acidic trachytes (together with rhyolites, phonolites, etc.) to basic types, i.e. basalts. The tuffs often carry impressions of the vegetation of their time from which their geological age can be determined.

Chapter 13

Geologic Maps

The contrast between a lava flow of black basalt and the prevailing light-colored granodiorite of the Sierra Nevada, or, half way around the earth, between a similar basalt flow and the gleaming white coral rock of Samoa. To show the outlines of such a flow on a map we must trace the contacts between the flow and the other rock.

This is done by plotting the contacts of the flow on the map in their proper relation to valleys and hills, the contour lines and by other features on the map such as streams and roads. The result is a *geologic map*. A geologic map shows the distribution of the different rock masses that underlie the ground surface plotted accurately to scale in relation to the topographic features and other control points on the map.

Such maps are economically useful. If basalt is needed to surface a road in the Sierra Nevada, or to build a breakwater in Samoa, we know from our geologic map where it can be quarried. From the contours we can also estimate the thickness of the flow, and thus calculate the approximate tonnage of basalt. From geologic maps, in conjunction with topographic maps, we can determine much of the size and shape of rock bodies hundreds or thousands of feet below the surface.

The economic power of this geologic tool is especially great in locating oil accumulations, water supplies, coal, iron ores, and other valuable substances hidden beneath a cover of soil and rocks. Though no surface indications of these valuable materials may be apparent, geologic mapping often reveals where they can be found by tunneling or drilling. The accuracy of such geological predictions has been proved again and again by actual tests.

The geologic map is also the principal tool in deciphering the history of a mountain range, the course of evolution of fossil organisms, the changes of local climates, or, indeed, the history of the earth itself.

Difficulties of Geologic Mapping

The distribution and relationships of rock bodies are not commonly so obvious as those in the High Sierra or on Samoan beaches. Indeed, in a fertile agricultural area like the lower Mississippi Valley, any systematic arrangement of the rocks is not readily seen. In such farming areas the rocks are nearly everywhere masked by soil or alluvial deposits.

Unweathered rock is exposed only in stream banks and deep ravines, or in road cuts, quarries, and other artificial excavations. To make a geologic map in such regions is not easy and often requires digging pits and trenches in critical spots, or drilling holes and examining fragments of the rocks penetrated by the drill.

EARLY GEOLOGIC MAPS

The principles of geologic mapping were developed chiefly in western Europe, where the rocks are seldom well exposed. Among the earliest geologic maps that showed the relations of sedimentary strata over a considerable area are two maps of the region around Paris, France, published jointly by the French naturalists Georges Cuvier and Alexandre Brongniart in 1810 and 1822. At about the same time (1815) William Smith, an English surveyor, published a geologic map of England which was one of the milestones in the development of geology.

Long before the work of Cuvier and Brongniart, French scientists knew that the rocks near Paris are gently tilted layers of limestone, clay, gypsum, and sandstone. These rocks could be seen in many natural and artificial exposures, notably in the pits dug by the makers of pottery and porcelain in their search for plastic clay. As early as 1782, Lavoisier took time from his epochal discoveries in chemistry to demonstrate that quarry after quarry near Paris showed the same succession of clay, limestone, gypsum, impure limestone, and sand, from the base upward, with a siliceous limestone forming a cap at the top.

Cuvier and Brongniart went much farther than Lavoisier. They demonstrated that certain characteristic fossils were present in some of the different layers, that some layers changed in character both laterally and vertically, and that the different strata near Paris could be grouped into units called *formations*, whose thickness and physical characteristics are sufficiently distinctive to allow them to be traced continuously for long distances.

Succession of the Rocks

East of Paris one enters a series of low hills and narrow lowlands that curve in a broad arc, partially encircling the city. These hills have steep slopes on the side away from Paris and more gentle slopes toward the city.

The lowland just beyond the first group of curving hills, though largely covered with soil, contains scattered rock *outcrops* (natural exposures) and also man-made excavations, all of which expose chalk, a variety of limestone.

The chalk occurs in a lowland that almost completely encircles Paris. On the side of the lowland farthest from Paris, the rock is gray, with many layers of pale green sandstone; closer to Paris the chalk is a white, very porous rock containing numerous potato-shaped nodules of black flint scattered along thin layers within it. Such layers may be separated from one another by 15 to 20 feet of massive chalk containing no flint nodules.

The layering in both the gray and white chalk inclines gently downward toward Paris; hence Cuvier and Brongniart concluded that the white chalk rests upon the gray. They were thus able to separate the chalk into two distinct mappable units which they called *formations*: the *Lower Chalk*, chiefly gray chalk and greensand; and the *Upper Chalk*, a massive white chalk with flint nodules. It is the Upper Chalk which appears across the English Channel in the "white cliffs of Dover."

Cuvier and Brongniart recorded more than fifty different kinds of fossil shells and other animal remains in the Upper Chalk, all of which resemble, although they are not identical with, those of animals now living in the sea.

If one climbs the steep slope of the hills on the Paris side of the chalk lowland, he finds scattered outcrops of a plastic clay in the bottoms of little ravines where the thin soil cover on the steep hillside has been washed away by the rains. This clay rests upon the chalk. Many pits show that near the base of the hills all the outcrops are of clay, but ravines farther up the slope show loosely consolidated sandstone resting on the clay. Still higher are exposures of limestone and marl (impure limestone). In all these beds the stratification is distinct, and each stratum slopes gently toward Paris.

The clay that overlies the chalk yielded no fossils, leading Cuvier and Brongniart to conclude that, although both chalk and clay were originally deposited in water, they were formed under different environmental conditions.

Apparently, the animals that flourished while the chalk was accumulating did not live in the muddy water from which the clay was deposited. In some places, at the contact between chalk and clay, fragments of the chalk were found included in the lowermost bed of clay. Cuvier and Brongniart concluded that the chalk must have been a coherent rock when the deposit of clay began to form, and hence that considerable time had elapsed between the accumulation of the chalk and that of the clay.

Eventually, Cuvier and Brongniart systematically worked out the succession of the formations all the way to Paris. It included several limestones, sands, clays, and gypsum beds. Some beds contained marine shells; others bones of land mammals and birds, skeletons of fresh-water fish, and impressions formed by the leaves of land plants. Each formation thus showed certain characteristic physical features and also contained distinctive fossils by which it could be recognized.

In its larger features the succession was constant throughout the Paris Basin, but not every bed, nor even a group of beds making up a formation, could be followed all the way around the periphery of the basin. In places, beds had been eroded away before the overlying beds were deposited; other beds had been deposited as lenses in discontinuous small basins, or were heaped up by currents in the water, as are present-day spits and sand bars.

Identifying Formations by Fossils

Such variations made Cuvier and Brongniart's task of mapping difficult. For example, the distinction of one limestone from another, among so many, could not always be made on the basis of colour, details of bedding, thickness, or other physical characteristics.

Not all these difficulties were solved by Cuvier and Brongniart; modern geologic maps of the Paris Basin show many refinements in the subdivision and mapping of the formations that escaped their pioneer studies. But the basic principles they established have served well in all later studies. Perhaps the most important conclusion they reached was:

Each formation (closely related group of strata) contains its own characteristic assemblage of fossils.

This generalization has proved to be of fundamental value in correlating isolated outcrops of strata, even across seas and oceans. It has been checked again and again by the Law of Superposition. For example, approximately the same changes in the fossils occur in the same order as one goes up through the strata of England, France, and North Africa.

William Smith's Geologic Map of England

The first geologic map of England, published in 1815 by another of Cuvier's contemporaries, an English surveyor named William Smith, was also a great stimulus toward the wider use of fossils in geologic mapping and correlation. Smith's map rep resented a considerable advance over the maps of the Paris Basin prepared by Cuvier and Brongniart. As might be expected of a surveyor, Smith took pains to locate the contacts of the rock formations accurately and to plot them correctly on the map with reference to streams, roads, and other control points. Starting with the exposure of a contact on a canal or river bank, he would follow its approximate position across a soil-covered hill, using as guides the small rock fragments in the soil or at the mouths of fox or rabbit holes, until he could again locate the contact exactly in another valley where it was exposed in the bed of a stream.

Smith, like the French geologists, learned much about the succession of strata by studying man-made excavations. He was employed as a surveyor on many canals built just prior to 1800. On the Somersetshire Coal Canal, for example, Smith superintended the excavation work for more than six years, necessarily observing the details of the strata in several miles of cut.

He noted that in many excavations the strata are not horizontal, but are inclined to the horizon at an angle. He also records that "each stratum contained organized fossils peculiar to itself, and might, in cases otherwise doubtful, be recognized and discriminated from others like it by examination of them."

Employed on many engineering projects throughout England, Smith often traveled as much as 10,000 miles in a year. On all these travels he made full notes of the succession and nature of the rocks in different parts of the country. Finally, after twenty-four years of observation, he published his colored geologic map of England which has taken its place among the great classics of geology.

Smith's map profoundly influenced the development of geology. A recital in words of the results of his twenty-four years of labour would be too bulky for use, but the principal relations that he found could be quickly grasped by any trained observer once they were plotted on a map. Scientists saw that the answers to many problems could be had by carefully plotting the position of the rock masses that make up the earth's crust. Smith's skill in predicting the kind of rock that would be found in an excavation or tunnel, or the amount of overburden that lay above a concealed coal or clay bed, led men to think of the economic aspects of geology and of the practical ways in which a knowledge of the succession of rock strata may be used.

Perhaps Smith's most significant advance, however, was to determine the *stratigraphy, the order of succession,* of the different sedimentary formations for an entire country, and thus to prove the continuity of individual formations over great areas.

He showed conclusively that, if a given bed or stratum occurs above another distinctive formation in one locality, it occurs nowhere else below it. Indeed, the succession of strata that he established was found to hold not only for England, but also for much of Europe. Eventually, the subdivisions that he and other pioneers established were extended and refined into the *Standard Geologic Column* to which we now refer sedimentary formations throughout the world.

METHODS AND PRINCIPLES OF GEOLOGIC MAPPING

This brief account of the early geologic maps gives us glimpses and hints of the methods and principles used in geologic mapping. We should now summarize these principles, consider their validity, and show how they are applied in actual field work.

On William Smith's geologic map of England, lines representing the contacts between different rock formations are drawn for distances equivalent to hundreds of miles on the ground. Yet, in tracing an individual contact for 100 miles, Smith probably found, on the average, less than 50 exposures where the actual contact could be seen on a clean rock face. How, then, can his map record the real distribution of the rocks? Can it be anything but a guess? The eyes of a geologist are no more capable of seeing the bedrock through a cover of soil than those of any other observer. How, then, can the geologist make inferences about the position of the bedrock underground that will withstand objective tests, such as those provided by digging wells and sinking mine shafts?

The succession must be pieced together from scattered outcrops. Although in an area of a few square miles a geologist may find only one or two outcrops in which he can see the actual contact between two formations, he will doubtless find 100 or more outcrops composed entirely of rock belonging to one or the other formation. He thus has some clue as to the position of the contact with respect to each outcrop.

The problem is roughly analogous to drawing a contour line to conform to the elevations determined at 100 or more control points. To use scattered rock outcrops, however, requires that the individual formations be correctly identified (correlated) from outcrop to outcrop, and this is not always easy.

Four Fundamental Postulates of Geologic Mapping

There are four fundamental postulates that underlie the making of a geologic map. The third is merely a commonsense deduction from these two, and, like them, was first stated by Steno: A water-laid stratum, at the time it was formed, must continue laterally in all directions until it thins out as a result of non-deposition, or until it abuts against the edge of the original basin of deposition. This is the Law of Original Continuity.

An important corollary of this law, not fully appreciated by Steno in 1669, but well known to the geologists of France and England at the beginning of the Nineteenth Century, may be stated: A stratum which ends abruptly at some point other than the edge of the basin in which it was deposited, must have had its original continuation removed by erosion, or else displaced by a fracture in the earth's crust. These four postulates—

- Superposition (the higher bed is the younger).
- Original horizontality (stratification planes are formed roughly parallel to the earth's surface).
- Original continuity.
- Truncation by erosion or dislocation—are the basis for many of our interpretations of the relative relations of strata.

They are not absolute rules that can be rigidly applied. For instance, some beds, once horizontal, have been highly tilted and even overturned by movements of the earth's crust, so that a stratum formerly beneath another may now lie inverted on it; other strata, as at the front of a steep delta, may have been deposited on appreciable slopes; the edges of many landslides end abruptly instead of thinning to a feather edge.

Such exceptions, however, are not common and can generally be recognized easily by the geologist. How do we correlate between scattered outcrops when we are in the field making a geologic map?

CORRELATION OF ROCK OUTCROPS

In a ravine on a grassy hillside we may see a bed of clay with well-marked horizontal stratification; we assume it continues horizontally into the hill at the same elevation, for how else can its horizontal stratification be extended? If we go 200 yards along the same contour, without seeing an outcrop, and then find a clay bed at the same elevation in another ravine, we may suspect it is the same bed. If both are gray, the probability is heightened.

If both show lines of nodular lumps (concretions) along the stratification planes, and if the size and spacing of these lumps is about the same, we are still more confident of our correlation. If both rest on red limestone, both are overlain by fine-grained brown sandstone, and both contain the same kinds of fossils, we become practically certain of the correlation. We are now justified in assuming that the bed is continuous beneath the soil between the two exposures. If we go on a little farther and come to a deep ravine in the hill which shows a very large outcrop containing not a few feet of strata but several hundred, and if in this section there is only one clay bed with features identical to those we saw in the two small outcrops, we have still further evidence that our mapping is correct.

We can now map the clay, for the contacts of the bed with those above and below will parallel the contours, as they are horizontal planes along the top and bottom of the clay bed. We have, furthermore, gained information on the thickness and relations of the red limestone below and the brown sandstone above the clay bed, because in this larger outcrop they are exposed through a much greater thickness. For example, we may find that the brown sandstone is 180 feet thick and is overlain by a thick bed of distinctive black limestone crowded with fossils.

If, now, we go on to still another outcrop and find here the clay bed is a little thinner than in the last outcrop, if instead of resting on red limestone it now rests on a pink limy sandstone, and if it has a pebbly green sandstone instead of the fine-grained brown sandstone above it, we would be less certain of our correlation, though we would not regard it as impossible.

To check on the correlation, we might go to a lower elevation and study the relations of the sandstone in various outcrops. If we found it gradually varying from limestone to sandy limestone and then to pink sandstone, our correlation would be strengthened. It would be further confirmed if we walked up the hill and found that the pebbly green sandstone above the clay is overlain by a black limestone full of fossils exactly like those in the limestone that overlay the brown sandstone in the large exposure.

Thus in geologic mapping there is invariably an element of judgment. Some correlations are certain, others are reasonably sure, and for still others there may be a reasonable doubt. As to the doubtful ones, two geologists may disagree, just as two equally qualified physicians sometimes disagree in the diagnosis of identical, but not sufficiently definitive, pathological symptoms. But nearly all such differences in interpretation have to do with minor features in a stratigraphic succession. Thicker groups of beds commonly exhibit enough peculiarities to lead any two careful observers to identical conclusions.

Geologic Sections

Geologic sections are commonly used along with geologic maps in all economic applications of geology. A geologic section represents the way the rocks would appear on the side of a trench cut vertically into the land surface. It resembles a geologic map in representing the projection of the rock boundaries to a surface, but this surface in a cross section is vertical and not the land surface of the earth.

As an example, let us make a geologic section of the horizontal clay bed just mentioned, taking our data from a geologic map showing the relations of the clay in a deep branching ravine where the beds are well exposed. The completed section appears below the geologic map, together with the construction lines used in drawing it. The geologic section portrays the strata as they would appear on the wall of a trench dug along the line A-B of the geologic map.

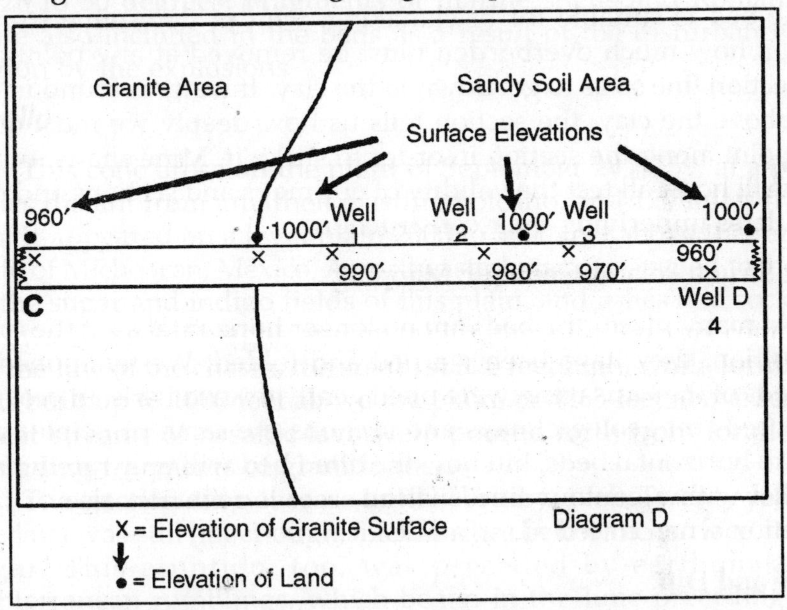

Fig. Construction of a Geologic Section from a Geologic Map

The section was readily drawn by making use of the contour lines on the map and relating the outcrops to them. First, it was decided to make both the horizontal and vertical scales of the section the same as that of the map, in order to avoid distortion. Then perpendicular lines were dropped from A and B (by means of a right-angle triangle) to the place below the map chosen for drawing the section.

Then a vertical scale showing the range in elevation from 1120 to 1,260 feet above sea level was laid off on the line projected from B. Horizontal lines were drawn across the section, from the point representing the elevation of the respective contours from 1,140 to 1,260. Then a perpendicular was dropped from each point of intersection of line A-13 with a contour line on the map, to the horizontal line on the section representing the appropriate contours. Thus, points were plotted on the section to guide the sketching of the *surface profile,* the irregular line A2 -B2 of the section. Now perpendiculars could be dropped to the profile from the intersection of the map line A-B with each contact of the clay stratum.

There are five such intersections shown along the line of section—three on the top of the clay stratum, two on its base. Horizontal lines drawn through the projections of these points on the surface profile complete the cross section and show the relations of the clay bed to the underlying limestone and overlying sandstone.

Many practical uses can be made of geologic sections. If the clay in the bed is suitable for brickmaking, we can determine from the section how much overburden must be removed at any point along the section line so as to get down to the clay. If water is found in wells just above the clay, the section tells us how deeply we must drill at any point along the section in order to strike it. Mine shafts, tunnels, and drill holes all test the validity of our maps and sections and of the postulates underlying their construction.

Rock Structure and Geologic Mapping

At many places the beds are no longer horizontal as at the time of deposition; they have been warped and folded. We mentioned that the beds of the Paris Basin were tilted gently toward Paris. In projecting tilted beds on geologic maps and sections the same principles apply as with horizontal beds, but now the tilted bed will only rarely project parallel with a contour line. Instead, it will generally rise or fall in elevation when traced along a hillside.

Strike and Dip

To project a tilted stratification plane either along its surface outcrop or into the ground, the first step is to determine the *strike and dip* of the stratum. If a tilted stratum intersects the surface of a lake the trend of the intersection of the water's edge with the stratum —that is, the bearing, or compass direction, of the line of intersection of the lake surface (a horizontal plane) with the stratification (a tilted plane)-forms a definite line of reference. This trend, or compass direction, is called the *strike* of the stratum. We know that the stratum will project along it in the horizontal direction unless terminated by erosion or other causes.

All horizontal lines drawn on the stratum will have the same trend as the line marking the edge of the lake against it; *each is a strike line.* Strike is measured in the field by sighting along an inclined stratum with a compass equipped with a level bubble so that it can be held horizontally. The direction of strike is recorded from the north. Thus a bed which intersects a lake surface along a line running exactly northwest-southeast is recorded as having a strike of N45°W (which is, of course, the same as S45°E, or northwest-southeast).

The dip of a bed is the maximum acute angle between a horizontal surface and the stratification plane. To go back to our example of the lake against an inclined stratum, the dip is the angle between the surface of the lake and the submerged continuation of the tilted bed, measured at right angles to the strike. Dip angles are measured by a *clinometer,* which is simply a level bubble attached to an arm inside the compass that swings against a scale graduated in degrees.

The *direction of dip* must not be confused with the *angle of dip.* The direction of dip means the compass direction toward which the tilted bed is inclined down into the earth; in our example, it is the direction (not the inclination) taken by a drop of water lifted out of the lake and allowed to run down the surface of the inclined bed. *The direction of dip is always at right angles to the direction of strike.*

TOPOGRAPHY AND GEOLOGIC MAPPING

A dipping bed can be projected along a contour only when the trend of the hillslope is parallel to the strike. If the strike does not parallel the contours, the trace of the bed along the hill will lie either higher or lower than the first outcrop, depending on whether the bed is followed in the up-dip or down-dip direction.

This simple geometrical fact is the basis for most determinations of attitude of beds with low dips, for it is difficult to read a clinometer more closely than 1°. It is also the basic clue to reading a geologic map in order to glean from it the succession of the beds and their structure.

In the Paris Basin most of the streams drain to the Seine. East of Paris they flow westward, cutting approximately at right angles through the arcuate ridges. Applying the relation between the dip of a bed and its trace on the land surface, we find that, if we select some stratum such as one of the thin layers of greensand in the Lower Chalk, or one of the thin limestones that lies above the Plastic Clay, and trace the outcrops of this bed westward along the banks of a stream, the bed decreases gradually in elevation when followed to the west until it reaches the level of the stream; thereupon it crosses the stream bed, reverses its direction and climbs higher and higher when followed eastward on the opposite bank of the stream. In other words, the outcrop pattern is a "V" with the point of the "V" directed downstream. This can only be explained by the fact that the bed is inclined westward, and that its dip angle is greater than the fall (gradient) of the stream. The fact that the Upper Chalk crops out further west in the valley of the Marne and Aisne rivers than it does on the intervening divides proves that the contact of the chalk with the overlying Plastic Clay also dips to the west. By using such facts the east-west section forming the lower edge (bottom) was drawn, although each formation shown on this section is projected to much greater depths within the earth than we can actually observe in the field. The basic principle of original continuity is the same as for horizontal beds, but in projecting the bed to a geologic section, account must be taken of the dip of the beds.

RELATION OF TOPOGRAPHY TO UNDERLYING ROCKS

There is generally a relationship, faithful in areas of little soil but more generalized in areas of deeper soil, between the surface form of a country and the underlying bedrock. In the Paris Basin the curving ridges mark outcrops of resistant strata such as sandstones. Porous chalk layers and other easily eroded rocks are found in the lowlands.

Weathering and erosion have etched the surface into relief, forming lowlands on the less resistant beds and leaving the more resistant ones standing as hills. The curving pattern of these hills and lowlands shows that the rocks of the Paris Basin have been warped into a shallow saucerlike basin. Indeed, the whole succession of strata resembles a pile of saucers of diminishing size, with Paris near the centre of the smallest and uppermost saucer.

The correlation between topography and structure in the Paris Basin is by no means perfect—much closer adjustment is noted in deserts where the soil cover is thin or absent. However, the topography usually indicates the general trend of the underlying rocks. Generally, sandstone is more resistant to erosion than shale and hence forms a ridge. It is also more permeable to rain so that, if it rests on shale, there is likely to be a line of springs or a strip of flourishing plants along its lower contact.

Thus, the minor topographic features of an area nearly always furnish clues to the characters and trend of the underlying bedrock. If we find conglomerate on top of a low ridge at one point, it suggests that conglomerate may underlie the ridge throughout its length. This is easily checked by seeking other outcrops along its summit. A road may expose chalk in a lowland. Perhaps a well has been dug in the same lowland a thousand feet away. Does it show chalk on its walls? By piecing together such information, the geologist can generally establish the stratigraphic succession, discover the variations in thickness of beds, and ultimately plot their positions on a map.

Even in country with a thick soil it is still possible to apply the principles of geologic mapping and correlation by using data gained from excavations, well borings, or geophysical methods.

Limitations of Scale

As in all other maps, geologic maps require the rigorous selection of data; they emphasize some features at the expense of others. The smaller the scale of the map, the fewer the details that can be shown. On which shows only a few square miles of the Paris Basin near Epernay, details of thin layers above the chalk are shown that could not possibly be indicated on the small-scale map of the entire Paris Basin prepared by Cuvier and Brongniart.

Thus the geologist must select his *geologic map units,* or *formations,* to fit the scale of the map. The larger the scale and the better the exposures, the larger the number of formations that can be shown within a given area, although the number per square inch of map remains about the same.

Formations

The basic unit of the geologic map is the formation. There are two criteria for selecting a formation: first, its contacts (i.e., the top and the bottom of a sedimentary formation) must be recognizable and capable of being traced in the field, and second, it must be large enough to be shown on the map. In their studies of the Paris Basin, Cuvier and Brongniart noticed faint stratification surfaces within the Lower Chalk. The chalk above and below these stratification surfaces, however, was so nearly identical in appearance and fossil content that such subdivisions of the strata were not regarded by the two geologists as either significant or capable of being successfully traced and mapped in the field. On the other hand, the contact between the Upper Chalk and the overlying Plastic Clay was mapped as a formation boundary because of the marked contrast in the rocks. A still more cogent reason for selecting this contact to map was the evidence that it represented a considerable interval of geologic time—enough for the underlying chalk to have become firmly coherent, so that it could be incorporated as pebbles and fragments in the lowermost bed of clay. The stratification within the clay was more prominent than the chalk, but again the similarity of each bed of clay to its neighbors made it difficult, if not impossible, to trace any particular stratification plane far.

The beds above the Plastic Clay posed a somewhat different problem to Cuvier and Brongniart. Here, there were many different kinds of rock in relatively thin layers—limestone, shales, sandstones, beds of gypsum, and clay. At some outcrops an individual bed, perhaps a clay only a foot thick, could be seen to thin and "lens out" within a few tens of feet. Or the overlying sandstone, perhaps 15 feet thick in this outcrop, could be seen to thicken when followed across country in successive outcrops, and then to thin again and perhaps ultimately disappear. Only on a very large scale could each thin layer be shown on a map, and it would take a prodigious amount of time to trace the contacts. Such a series of thin, variable beds, often including very diverse rock types, were grouped together by Cuvier and Brongniart as a single formation. Although the thin beds of such variable formations may be individually lenticular, and indistinguishable either from other beds of the same formation or adjacent ones, nevertheless the group as a whole is recognizably different from formations above and below it.

Depending on the scale of the map, the abundance of exposures, the character of the beds, and, not least, the discrimination of the geologist, very different map units (formations) might be selected in a certain section. Any differences are adequate to justify calling a particular bed, or any closely related group of beds, a formation, provided the differences allow the formation to be recognized in scattered outcrops, and provided the top and bottom of the formation can be traced in the field. The purpose of the map and its scale largely determine what units to select as formations.

Geologic formations in the United States are nearly always given a place name from some geographic locality near which the formation was first identified. This is followed by the name of the *dominant* rock variety composing it, or, if it is composed of a great variety of rocks, by the word "formation." Thus: the *Austin chalk*, the *Columbia-River basalt*, the *Chattanooga shale*, the *Denver formation*. In Europe the practice is less formal; many formations are named from some characteristic fossil (the *Lingula flags*, a thin-bedded sandstone containing abundant fossils of the genus *Lingula*), some economic character (*Millstone grit*), or even a folk name (*Norwich Crag*).

PALEONTOLOGY: THE USE OF FOSSILS IN GEOLOGY

Paleontology deals with the life of the geologic past. It includes the identification and classification of fossil animals and plants, and the study of their interrelationships and development.

Correlation and Faunal Succession

At first, Cuvier, Brongniart, and Smith used fossils to identify the beds in different outcrops very much as one might use unusual kinds of pebbles or chert nodules. But much more than this developed from their work. That many fossil shells found far inland are the remains of marine animals had been recognized for centuries. The great advance made by Cuvier and Brongniart was to recognize that there were systematic differences in the fossils in the strata of the Paris Basin.

When they arranged the fossils collected from the various strata in the same order as that of the beds in which the fossils were found, they noticed that the fossil sequences differed from one group of beds to another, and that these differences were systematic: The fossils from the lower (older) beds were not so similar to animals now living as were those from the higher (younger) beds. No matter whether the particular fossil was a clam, a sea snail, or any other kind of organism, this was broadly true.

Geologic Chronology

This discovery was one of the most significant in the history of geology. Now fossils became recognized as diagnostic markers of a specific period of time during which the bed containing them was deposited. From studies of thousands of different species of fossils it has been discovered that some shells, even from very ancient rocks, differ little from shells of living organisms, but that such stable and long-continuing forms are very few.

Although a rare long-ranging species may persist through a thick series of strata, its associates generally change gradually from bed to bed until none of its original companions in the lower part of the series are present in the higher. Conversely, an abundant species may become absent in higher strata, whereas its associates persist. Thus individual species range through different thicknesses of beds, which must mean different durations of time.

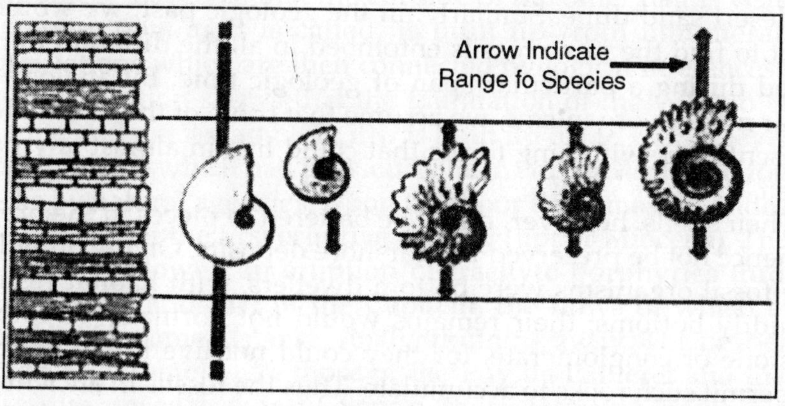

Fig. Using Fossils to Date Beds

Here, the key was found to much of earth history. Once the systematic changes of fossils wihin the sequence of beds of the Paris and London areas were recognized, it became a challenge to workers to discover whether similar changes in the succession of fossils could be found elsewhere. Within a generation, parallel successions of fossils became known in the sequence of strata from many lands. Such successions, partly identical and partly supplementing one another, furnish the basis for geologic chronology.

It is true that our geologic calendar built on fossil successions is not calibrated in years. Rough dates in years, however, are gradually being found from studies of radioactive minerals. Few fossil forms have a truly cosmopolitan distribution, but there are some forms, especially among the free-swimming organisms, that are nearly world-wide in distribution. Wherever such wide-ranging forms have been found in the fossil record—whether in Europe, Africa, Asia, or America-the sequences of rock strata are consistent, one with another. Innumerable such examples justify the inference that: Like assemblages of fossil organisms indicate similar geologic ages for the rocks that contain them.

Sedimentary Facies and Facies Fossils

In correlating rock strata by means of fossils it is important to keep in mind the limitations to the spread of organisms imposed by their natural habitats. Many different depositional environments exist: floodplains of rivers,. estuaries protected from the open sea, ocean beaches, coral reefs, and many others.

Each has a characteristic group of animals and plants living contemporaneously with the somewhat different but equally characteristic groups of living organisms found in the others. For example, one does not expect to find antelopes in a coral reef, nor corals in a desert sand dune. Similarly, in the geologic past, we would not expect to find the same fossils entombed in all the different deposits formed during a particular span of geologic time. By analogy with similar modern organisms, we assume that some of the marine fossils represent free-swimming forms that could live in almost any part of the sea.

Their shells, however, might be battered to pieces in the breakers and hence not be preserved in nearshore deposits. On the other hand, some fossil organisms were bottom dwellers, with habitat restricted to muddy bottoms; their remains would not normally be found in limestone or conglomerate, for they could not live in environments that permit such rocks to accumulate. Thus the fossils in ancient rocks of even the same age may differ with variations in the conditions of deposition of the rocks. Rocks of the same time span that were formed in different depositional environments are said to represent different "*sedimentary facies*."

Most fossil organisms were restricted in their life habitats, and hence tend to occur typically in rocks of a definite kind of sedimentary facies. The term *facies fossils* is generally used for groups of fossil organisms limited to particular kinds of sediments. Thus we speak of the "limestone reef facies," which may give way along the strike of the beds to a mud-bottom-dwelling shale facies."

One relationship that may occasionally be found between two facies fossils. A long-range fossil (Fossil A) occurs above another shorter-ranging kind (Fossil B) in one region, but below it in another. Such reversals are generally the result of differences in environment of deposition of the two fossils, but they may occur simply because of the greater thickness of rocks during one fossil's time range, or even because of accidents of preservation.

Likewise, it can be seen how a particular fossil may be absent from a certain bed containing a second kind, though it is found both below and above that bed. The environmental conditions (sedimentary facies) changed during the life-span of the longer-ranging form, excluding it temporarily from the area of deposition, but allowing it to return when the life habitat again became favorable.

Despite these complexities—and there are still others— Cuvier and Brongniart's generalization has been abundantly justified: The older the rocks, the less their fossils resemble living forms; the younger the rocks, the more their fossils resemble living forms, particularly those found in the same depositional environment.

THE GEOLOGIC COLUMN

The *standard geologic column* is the basis for the geologic time scale. The standard column is composed of a sequence of European formations because it was developed by expanding the stratigraphic sequences worked out by Smith, Cuvier, and Brongniart. Its fossil assemblages form the standard of comparison for sequences in all parts of the world. Correlations from one part of a continent to a distant part, or even to another continent, are based on assemblages of fossil plants and animals.

Such correlations are sometimes handicapped by the local, rather than cosmopolitan, distribution of many fossils and by differing conditions of deposition. On the other hand, they are facilitated by the vast wealth of former living things; thousands upon thousands of varied forms are preserved in the fossil record. The *standard time scale* is a scale representing the periods of deposition of the rock units in the standard stratigraphic column.

Simple Early Column

As early as the middle of the Eighteenth Century, Italian and German geologists had classed their local rocks into three groups: Primary (rocks like granite and gneiss, with neither bedding nor fossils), Secondary (cemented sedimentary rocks, generally exposed in the mountains), and Tertiary (weakly consolidated sedimentary rocks of the lowlands, which rested upon the Secondary rocks). Though fossils were known from both Secondary and Tertiary rocks, distinctions among them from bed to bed were either not made or not used in the Eighteenth Century classification.

Standard Column

The present standard column was pieced together in the Nineteenth Century from strata exposed in various parts of Europe. The major subdivisions of the column are called *rock systems*. Most of the systems are represented, at least in part, in the area of Britain mapped by Smith and his early followers.

Smith grouped the oldest strata in his section as the *Old Red Sandstone*. These rocks contain few fossils and, in fact, are largely land-laid beds, but to the south they interfinger with fossiliferous marine beds widely represented in Devonshire. The Old Red Sandstone is, therefore, now considered a part of the Devonian System. Resting on the Old Red Sandstone is a group of strata called by Smith the *Mountain Limestone*, and this series is in turn overlain by a succession of sandstones, shales, iron ores, and coals which Smith called the *Coal Measures*. These rocks and the underlying Mountain Limestone are now grouped together as the *Carboniferous System*, named from the abundance of coal in the rocks.

Above the Coal Measures Smith recognized a series of beds called the *Magnesian Limestone*, followed in turn by the *New Red Sandstone*. Both these formations are poorly fossiliferous and largely of continental rather than marine origin. The Magnesian Limestone is now considered part of the *Permian System*, named from a province in Russia just west of the Ural Mountains, where marine fossiliferous rocks occur in the same stratigraphic position. The New Red Sandstone, on similar grounds, is now regarded as part of the *Triassic System*, named from the sandwich of a marine limestone between two continental formations found just above the Permian System in Germany.

Overlying the New Red Sandstone, Smith recognized a whole

series of richly fossiliferous beds which he divided into many formations, partly because the abundant fossils facilitated correlations and partly because the canals he constructed lay across these beds, and he thus had ample time to work out the details of their successions. Some years later many identical fossils were found in the rocks of the Jura Mountains of Switzerland and France. The name *Jurassic System,* to which these strata are now referred, recalls this fact. Jurassic rocks also underlie the Lower Chalk of the Paris Basin.

Toward London, the Jurassic strata are overlain by greensand, and this in turn by chalk. This series is grouped as the *Cretaceous System* (named from the Greek word for chalk). The fossils are like those of the chalk in the Paris Basin. Overlying the chalk, Smith found the *London Clay,* a group of beds rich in fossils that are like the fossils in the beds above the chalk of the Paris Basin. More than a generation after Smith, the stratigraphy of the highly disturbed rocks below his Old Red Sandstone was worked out. Three new systems were recognized: in descending order, *Silurian, Ordovician* (both names of old British tribes), and *Cambrian,* from the Latin name for Wales.

The Cambrian strata are the oldest rocks that contain fossils in any abundance, although sedimentary rocks thousands of feet thick are found beneath them. In this book such older rocks are grouped as the *pre-Cambrian.* They have great bulk and complexity and can be classified in various ways, in spite of the absence of fossils. Perhaps refinements in the dating of rocks by radioactivity may some day enable us to classify the pre-Cambrian rocks in an orderly chronological sequence, just as fossils have permitted us to do for the Cambrian and younger systems.

GEOLOGIC TIME SCALE

The standard geologic column is the basis for the *geologic time scale.* The names applied to the systems of rocks are also used for the *periods of time* during which the respective systems of rocks were deposited. Thus, we use the term *Carboniferous Period* for the time during which the *Carboniferous System*—the Mountain Limestone and Coal Measures of Smith—were laid down. Appended to it are remarks as to the sources of the names and the ages of several minerals as found from studies of radioactivity. From these ages in years it is apparent that more subdivisions are recognized among younger rocks than among older. Just as in human history, the documents are more numerous and the gaps in knowledge less wide the more recent the period with which we deal.

Gaps in the Time Scale

Divisions of the standard column are based on abrupt changes in the fossil assemblages of the strata in Europe. These points of division naturally were those marking the longer intervals of erosion or nondeposition in the European stratigraphic series. The longer the interval lost by erosion, the more prominent the fossil differences would be. As stratigraphic work was extended to other continents, however, fossil assemblages intermediate between those of adjacent periods as recognized in western Europe were discovered.

With more and more stratigraphic work the gaps in the type sections have continued to decrease. Beds containing fossil assemblages that pose "boundary problems"—that is, uncertainty as to their correlation with the upper part of one European system or the lower part of the next higher—are thus present at one place or another for nearly every systemic boundary.

Even the boundary between the Paleozoic and Mesozoic Eras, long thought to represent an interval during which none of the present continents received deposits, is apparently bridged by an essentially complete succession of beds in the Himalayan Mountains. Even in Europe, where the divisions were first established, further research has served to narrow the lost gaps in the record.

For example, a new division, the Paleocene, has been recently established for strata intermediate in age between Eocene and Cretaceous, where a long gap had been thought to exist. Now we find that different paleontologists refer the same strata (let us say, in Wyoming) to the Cretaceous or to the Paleocene, depending on their respective appraisals of the relative affinities of the fossils of these beds to that of one or another of the European sections.

Intercontinental Correlation

Thus, although the original divisions of the type section were based on stratigraphic relations of the beds, our correlations of distant strata are necessarily based on a *comparison of fossils*, not strata. The demonstration of an interruption in stratigraphic succession in the type area can give no help in correlating beds in a distant part of the world. A stratigraphic break in Nevada, for example, cannot be used as evidence that the beds above it are Permian and those below Carboniferous. This assignment of age to the Nevada strata can be made only on the basis of comparison of fossils with those of the type section of Europe.

The uncertainties posed by "boundary problems," however, are concerned largely with form instead of fact. Such disputed strata, it is generally agreed, must represent times between, or at least partially bridging, the gaps in the type areas of Europe. Further detailed stratigraphic work cannot but fill other parts of the gaps. This is indeed a tribute to the precision of correlation by fossils rather than a criticism of its uncertainties.

THE DURATION OF GEOLOGIC TIME

With the building of the geologic time scale there came a growing appreciation of the vast duration of geologic time. Contemplation of even the time required to lay down the 500-foot thickness of chalk in the Paris Basin is upsetting to the man who thinks of time only in terms of the span of human life.

Some strata of the chalk are composed of the skeletons of minute animals and plants. Similar deposits are accumulating today at rates so low as to defy measurement—certainly no more than a few inches per century and probably much less. Yet the chalk represents only a small part of Cretaceous time, and the Cretaceous itself is a mere fraction of geologic time.

Not only the vast thickness of sediment impressed geologists with the immensity of geologic time, but also the vast parade of life recorded by the fossils—the development of thousands upon thousands of new species with advancing time and the gradual dying out of whole fossil assemblages and their succession by others—time after time—can only be conceived of as involving millions of years, unless greatly different rates of change prevailed in the past.

TIME REPRESENTED BY THE GEOLOGIC COLUMN

Spurred by their curiosity to translate geologic time into years, geologists tried many approaches. The simplest and most obvious one was to add the greatest thickness of strata of each period and thereby get the total thickness for all geologic time. This total, divided by the present annual rate of sedimentation, might be thought to give the length of geologic time in years.

However, so many assumptions are involved that such a calculation becomes meaningless. How can we arrive at an average annual rate of sedimentation when we know that it may take a century to lay down less than an inch of chalk, whereas a desert cloudburst may deposit 40 feet of gravel, sand, and boulders overnight? And how can we measure the time of scour and nondeposition in marine sediments when we know that a single storm may remove the accumulation of years from part of the sea bottom and perhaps pile up more sand on a nearby beach than was deposited during the whole preceding decade?

In short, the rate of sedimentation varies so greatly and has been so little measured that the average annual rate can only be guessed at. This method can only yield an order of magnitude for geologic time. By such a calculation, the British geologist Sollas estimated (1899) that 34 to 75 million years have passed since the beginning of the Paleozoic —the larger figure including a guess at the length of the lost intervals in the record. These figures have a certain interest but surely are not reliable.

AGE OF THE OCEAN

The Irish scientist Joly made a more ingenious approach. He reasoned that the salt in the sea must have been brought in by the rivers, after being weathered from the rocks. Most of it has accumulated in the sea, as little is blown out by the wind or deposited by evaporation of sea water in desert lagoons or on tide flats. Hence the oceans must be growing saltier; and, if the amount of salt they now contain could be measured and divided by the annual increment, their age might be obtained. From the many elements in solution in the oceans, Joly selected sodium as the constituent best suited to this method.

The volume of the sea has been roughly computed from its average depth and area, and its composition is known from thousands of analyses. Thus the total sodium it contains can be roughly computed. Similarly, we have thousands of analyses of river waters, and from stream gages we know fairly well how much river water flows annually to the sea. From these figures Joly obtained a rough estimate of the annual increment of sodium. Dividing the first amount by the second, he found:

Joly realized, of course, that this computation disregarded many factors. For one thing, river discharge was probably not constant throughout geologic time. Furthermore, not all the sodium in rivers is derived from weathering; much is from sewage and industrial waste, some is wind-blown from sea beaches, and much is leached from former marine sediments. Thus the present annual increment is higher than in the geologic past. How much higher cannot be ascertained.

Again, there are great deposits of rock salt among the stratified rocks, derived from evaporated sea water. If returned to the sea, they would increase the tonnage in the oceans. Again we have no way of getting an accurate correction. We have noticed how magnesium ion reacts with limy ooze to form dolomite. In a somewhat similar way, we now know that sodium ion reacts with clay minerals and is incorporated into marine muds and thus removed from solution. Finally, large amounts of sea water are mechanically enclosed in the pores of marine sedimentary rocks and the sodium so contained is then removed from the ocean. When all these corrections are considered, it is obvious that Joly's 99 million years must be multiplied many fold, but, as no figure can be made quantitative (that is, subject to careful measurement for purposes of verification), the results yield nothing but the fact that the ocean has existed for a vast span of time.

AGE BASED ON CLASSICAL PHYSICS

Another estimate of the duration of geologic time was made by Lord Kelvin, the famous English physicist of the late Nineteenth Century. In his day, before the discovery of radioactivity, atom-splitting and reconstitution, and transformation of mass to energy, Kelvin could imagine no source of energy for the stars except the contraction of the gases composing them.

Every astronomical body, including the earth, radiates heat into outer space, he knew, and, therefore, decided that each must be cooling and hence shrinking in size. He computed that, even if the earth had contracted from a size so great that it originally touched the sun, no reasonable assumptions as to original temperature could allow its age to exceed 400 million years. In fact, it was probably only about 20 million years old.

Lord Kelvin's dictum—despite his prestige as the leading physicist of his day—was rejected by geologists. Though the rates of sedimentation and of organic evolution cannot be measured, no geologist could bring himself to accept so short a time as Kelvin advocated. They regarded it as "unreasonably" short, despite his exact mathematical demonstration. We now know that his basic assumptions were the source of his error. The discovery of radioactivity and of other sources of energy not known to Kelvin demolished his basic assumption that all stellar energy arises from contraction.

The discovery of radioactivity also made it possible to establish some points on the geologic time scale in terms of years, and to show that all previous estimates of the duration of geologic time were far too short.

THE RADIOACTIVE CLOCK

When Becquerel discovered radioactivity (1898), he opened new vistas in every science. One of the many byproducts of this discovery was the determination of some geologic ages in years.

A few elements, among them uranium and thorium, disintegrate spontaneously into other elements. The atomic nuclei of such elements are intrinsically unstable and emit *alpha* particles and *beta* particles (electrons). Each nuclear emission, whether of an alpha or beta particle, transforms the atom into a different element. Starting with uranium 238 (U_{238}) there are 15 steps in this natural process: 8 alpha particles and 7 beta particles are emitted, yielding ultimately a stable form of lead (Pb_{206}) which does not disintegrate further.

The rate of spontaneous disintegration varies tremendously with different elements. It is expressed in terms of the element's "half-life," which is the time required for half its atoms to disintegrate. The half-life of some members of the U_{238} series is only a fraction of a second; for U_{238} itself it is millions of years.

The half-life period of the entire chain from U_{238} to Pb_{206} has been determined to be 7,600 million years. In other words, if we start with 1 gram of U_{238}, in 7,600 million years only 0.5 gram will be left, in another 7,600 million years there will be only 0.25 gram, and so on. The other half (or three-fourths) has changed into lead, helium ions, electrons, and very small amounts of intermediate elements in the series. In no experiment has the disintegration rate been changed by heat, pressure, state of chemical combination of the element, or time. The half-life of a radioactive element is thus considered a constant, that is, a fundamental property of the element.

Many minerals, most comparatively rare, contain appreciable amounts of uranium. Some of these are intimately associated with intrusive igneous rocks and have been segregated from the magma during the last stages of its consolidation by their occurrence in dikes and pods that grade into the intrusive rock. By analyzing such minerals and finding the ratio of uranium-derived lead to the uranium still remaining it is possible to find their age, and hence the time in years since the magma from which they crystallized was consolidated.

Such analyses are difficult. The proportions of the different lead isotopes must be determined so that uranium-derived lead (Pb_{206}) can be differentiated from any ordinary lead that may be present. Ordinary lead is a mixture of isotopes with an atomic weight of 207.21. Furthermore, all uraniumbearing minerals contain U_{238} as well as the more abundant U_{238}, and may contain thorium.

Each of these also disintegrates to yield a lead having a different atomic weight. These additional radioactive substances make it difficult to find the ratios accurately. Absolutely fresh unweathered minerals must be used, for solutions circulating long after the mineral was formed might leach out the lead and uranium at different rates and thus produce great errors in the calculation of the age. Though many determinations have been made, only a few meet the rigid requirements with regard to analytical procedure and freshness of material.

Thus we have a fairly reliable method for dating geologic time in years. Furthermore, for the rocks older than the Cambrian, it is the only method available for interregional correlation. The oldest mineral from the United States (1,600 million years) thus far determined comes from the Black Hills of South Dakota. It was taken from a segregation in a granite mass that cuts across schists. Therefore, it must be younger than the schists, which for many years had been considered late pre-Cambrian because they have not been severely metamorphosed.

The Grenville gneiss of the Adirondacks and southern Canada, on the other hand, is highly contorted and severely metamorphosed, so that it has been regarded as extremely old pre-Cambrian. It turns out from the lead ratios that this rock is little more than one-third as old as the Black Hills rock. Such evidence shows the impossibility of using the degree of deformation and metamorphism of the rocks as a basis of correlation over long distances.

Index